FLESH

OF MY

FLESH

FLESH

OF MY

FLESH

THE

ETHICS OF

CLONING HUMANS

A READER

EDITED BY

GREGORY E. PENCE

ROWMAN & LITTLEFIELD PUBLISHERS, INC.
Lanham • Boulder • New York • Oxford

ROWMAN & LITTLEFIELD PUBLISHERS, INC.

Published in the United States of America
by Rowman & Littlefield Publishers, Inc.
4720 Boston Way, Lanham, Maryland 20706

12 Hid's Copse Road
Cumnor Hill, Oxford OX2 9JJ, England

British Library Cataloguing in Publication Information Available

Library of Congress Cataloging-in-Publication Data

Flesh of my flesh : the ethics of cloning humans : a reader / edited
   by Gregory E. Pence.
      p.   cm.
   Includes bibliographical references and index.
      ISBN 0-8476-8981-6 (cloth : alk. paper).—ISBN 0-8476-8982-4
   (pbk. : alk. paper)
      1. Cloning—Moral and ethical aspects.   2. Medical genetics—Moral
   and ethical aspects.   3. Human reproductive technology—Moral and
   ethical aspects.   I. Pence, Gregory E.
   QH442.2.F585   1998
   174'.957—dc21   98-11287
                                                                          CIP

ISBN 0-8476-8981-6 (cloth : alk. paper)
ISBN 0-8476-8982-4 (pbk. : alk. paper)

Printed in the United States of America

Design by Deborah Clark

♾ ™ The paper used in this publication meets the minimum requirements of
American National Standard for Information Sciences—Permanence of Paper for
Printed Library Materials, ANSI Z39.48–1984.

CONTENTS

ACKNOWLEDGMENTS

THE EDITOR AND PUBLISHER THANK THE INDIVIDUALS AND ORGANI-
zations that have granted permission to reprint the following essays in this
volume.

Chapter 1: "Moving toward the Clonal Man: Is This What We Want?" by
James D. Watson originally appeared in *The Atlantic Monthly* in May 1971.
Reprinted by permission of the author.

Chapter 2: "Don't Worry: A Brain Still Can't Be Cloned" by George Johnson
originally appeared in the *New York Times*, 2 March 1997. Copyright © 1997
by The New York Times Company. Reprinted by permission.

Chapter 3: "The Wisdom of Repugnance" by Leon Kass originally appeared
in *The New Republic*, 2 June 1997. Copyright 1997 The New Republic, Inc.
Reprinted by permission.

Chapter 4: "Begetting and Cloning" by Gilbert Meilaender originally appeared in *First Things* 74 (June/July 1997). Reprinted by permission of *First
Things*.

Chapter 6: "Whose Self Is It, Anyway?" by Philip Kitcher originally appeared
in *The Sciences* 37, no. 5 (Sept./Oct. 1997). Reprinted by permission of *The
Sciences*. Individual subscriptions are $28 per year. Write to: The Sciences, 2
East 63rd Street, New York, NY 10021.

Chapter 9: "Dolly's Fashion and Louis's Passion" by Stephen Jay Gould originally appeared in *Natural History*, June 1997. Copyright the American Mu-
seum of Natural History 1997. Reprinted by permission of *Natural History*.

Chapter 10: "Clone Mammals . . . Clone Man?" by Axel Kahn originally appeared in *Nature* 385 (February 1997). Copyright © 1997 Macmillan Maga-
zines Ltd. Reprinted by permission.

Chapter 12: "The Confusion over Cloning" by R. C. Lewontin is reprinted
with permission from *The New York Review of Books*. Copyright © 1997
NYREV, Inc.

INTRODUCTION

BACKGROUND ON CLONING

On 24 February 1997, every paper in the world carried a front-page story about a lamb named "Dolly" that had been cloned by Ian Wilmut and his fellow scientists at the Roslin Institute near Edinburgh, Scotland. The lamb had actually been born many months before on 5 July 1996, but PPL Therapeutics (the for-profit funding corporation that backed Wilmut's research) waited for approval of a patent for its cloning techniques before announcing its achievement. It was immediately apparent that these techniques might be applied to humans.

Cloning is an ambiguous term, even in science. It may refer to molecular cloning, cellular cloning, embryo twinning, or somatic cell nuclear transfer (SCNT). In molecular cloning, strings of DNA containing genes are duplicated in a host bacterium. In cellular cloning, copies of a cell are made, resulting in what is called a "cell line," a very repeatable procedure where identical copies of the original cell can be grown indefinitely. In embryo twinning, an embryo that has already been formed by sexual reproduction is split into two identical halves. Theoretically, this process could continue indefinitely, but in practice, only a limited number of embryos can be twinned and retwinned. Somatic cell nuclear transfer is the process of taking the nucleus of an adult cell and implanting it in an egg cell where the nucleus has been removed; this process could be used to originate a human child. In a variant of this process called "fusion," a process which produced Dolly, donor cells are put next to an enucleated egg and fused with a tiny electric current. A blastocyst, a preembryo—preimplantation embryo—of about a hundred cells or less, starts to develop because the pulse that produces fusion also activates egg development.

But what does "cloning" actually mean? In 1993, American journalists and many bioethicists were breathless upon hearing reports that scientist Jerry N. Hall had cloned seventeen human embryos into forty-eight in order to increase the supply of embryos in fertility clinics.[1] But Hall had merely twinned undifferentiated embryos; he had not "cloned" them in the ordinary meaning of SCNT. Hall had merely introduced an electric spark that causes unspecialized embryos of a few cells to twin, a process that had been used for years in the livestock industry. Although this was a useful technique to help infertile couples who could not produce enough embryos, it was neither a breakthrough nor startling.

In 1987, Wilmut attended a scientific meeting in Ireland and overheard a story that encouraged him to experiment with cloning. He heard colleagues describe how Danish scientist Steen Willadsen, working at Grenada Genetics in Texas, had created a lamb clone by "enucleating" an egg—removing the nucleus from the egg—and fusing what was left with a cell from a growing preembryo.

Wilmut succeeded in creating the adult lamb Dolly from differentiated, specialized cells of her adult ancestor. This was an endeavor that few had even attempted—the assumption that this could not be accomplished was held so firmly that many promising biologists had fled old-fashioned embryology in favor of other fields such as molecular biology.

The breakthrough by Wilmut and his team involved timing the cycle of interphase of the cell cycle of the donor cell and recipient egg so that the two were not out of sync. In previous attempts to create fusion in this rapidly changing cellular environment, mitosis had gone wrong, producing broken chromosomes, thus defective embryos. Wilmut, however, stopped all activity in egg cells (made them dormant by not feeding them) and then fused the donor nucleus with the enucleated egg. In this way, he timed the cycles correctly and finally got a good fit.

At the first conference on the ethics of mammalian cloning in June 1997, Wilmut himself stressed several points for other researchers in cloning to consider.[2] First, he admitted that his present techniques are very inefficient: He used 277 sheep eggs to produce 29 embryos, for thirteen pregnancies. He got three near-births but produced only one live lamb. Until others repeat his work, we will not know whether he was very lucky or very unlucky with his initial results: the odds of a live lamb produced this way may be 1/100 or 1/10,000. Second, Wilmut's techniques have caused some examples of "large offspring syndrome," in which babies born are monstrously large and cannot be delivered by normal means. Wilmut and his colleagues do not yet know if this syndrome is caused by nuclear transfer, by something in the uterine environment, or by problems in the cell culture. Third, despite widespread fears about possible damage to genes in the embryo from SCNT, the lambs that died shortly after birth were chromosomally normal. Fourth, Wilmut suggested that we should avoid thinking of cells as "differentiated" or "undifferentiated," but instead think of differentiation as a continuum, where a cell's points on it are not forever fixed.

MEDICAL IMPLICATIONS

The discovery that cell differentiation is not an all-or-nothing process but a continuum and a process that can be reversed fuels optimism about treatments for neurological diseases. One researcher reported that neural growth factor can turn off tumor-differentiated cells in nine days. Such a technique could be used to create special neural cells from embryos to treat people with neurological diseases or, even better, to change diseased neural cells to healthy ones.[3]

Another possible use of Wilmut's research is to modify the gene in sheep that models cystic fibrosis in humans, and use such sheep to develop genetic therapies for this disease. (In a classy move, Wilmut sold the first wool shorn from Dolly to raise money for the care and treatment of children with cystic fibrosis.) Wilmut did not intend to be the first person to create a live-born mammal by somatic cell nuclear transfer from an adult, differentiated cell. Instead, the aim of his research was to create a new, reliable method of inserting a gene into every cell of a mammal (the old method rested on random chance, with very low probabilities of success). Five months after the announcement of Dolly's cloning, the team at the Roslin Institute announced that it had indeed accomplished this goal.[4] The first lamb with a human gene (for cystic fibrosis) was called "Polly" because she was a Poll Dorset Sheep. One molecular geneticist said of this achievement, "After Dolly, everyone would have predicted this, but they were saying it would happen in five or ten years."[5]

Wilmut argues that his techniques offer great promise for humans. Indeed, many scientists think that his real achievement may not be in cloning but in allowing us to understand and control cellular differentiation, to derive undifferentiated cells from differentiated cells, to understand how cells age, and to treat diseases caused by mitochondrial DNA. Wilmut's techniques should also help create genetically altered organs of pigs such that these new organs would have less chance of rejection in transplantation into dying humans. Similar modifications in livestock could also combat a disease called scrapie that affects the livestock industry. There is also the potential of fusion of a nucleus to an egg for cell-based therapy for some diseases. Overall, the most exciting prospect is to understand and direct the process by which the body forms specialized cells and to one day be able to direct undifferentiated cells to specialize under human choice.

GENETIC QUESTIONS

One aspect of the debate about human cloning requiring further study concerns the genetic contribution of the host egg, which is especially important in determining whether an individual originated by cloning would be an exact, genetic copy of his ancestor. Wilmut does not know whether the donor DNA for Dolly came from mammary tissue or from a relatively undifferentiated mammary stem cell.[6] Stem cells are thought to be more malleable and capable of assuming new functions than other, more rigidly-defined cells. For example, stem cells are important in bone marrow transplantation.

Scientists know that some DNA, especially mitochondrial DNA, would come from the host egg. Mitochondria are the organelles (tiny organs) that fuel cellular reactions. They provide energy by metabolizing glucose, which in turn produces ATP and NADPH+, products which drive cellular reactions. In mammals, mitochondria are inherited in the female cytoplasm (egg), which is separate from the nuclear genes from the female that are inherited in sexual reproduction. One egg cell contains hundreds of mitochondria, which are randomly distributed among the new cells being produced by the growing embryo. Much about the behavior of mitochondria is still unknown. Mitochondria carry genes, but not every cell has the same bit of DNA that is a gene. Because of aging of cells and environmental causes, mitochondrial DNA can change over many years. Such a mutation is thought to be a cause of some diseases such as Alzheimer's disease, Parkinson's disease, some forms of diabetes, and other more esoteric diseases. No one really knows how much mitochondrial DNA contributes to these diseases, however, in part because no one really knows how many diseases and disorders are solely caused by genes.

Both the nucleus inserted into the enucleated egg and the mitochondrial DNA may have been affected by an adult's lifetime exposure to the environment. Due to environmental effects, Dolly's cells may already be six years old. Will her cells prematurely age? Is Dolly only an old sheep in a young sheep's clothing? There was speculation that the udder cells of a ewe might be especially susceptible to environmental influences such as radiation and antibiotics and thus be different than the fresh cells of a baby lamb. When asked whether Dolly should be considered seven months old, which is how long she had been alive at the first press conference, or six years old, as a genetic replica of a six-year old sheep, a reporter wrote, "Dr. Wilmut's clear blue eyes clouded for a moment. 'I can't answer that,' he said. 'We just don't know.' "[7]

HUMAN CLONING: MORAL ISSUES

A convenient way to think about moral issues raised by human cloning is to divide them into those concerning the rightness of the act of originating a child by cloning and those involved in settling disputes when citizens cannot agree about the rightness of such an act. The first group of issues concerns content; the second concerns developing a process for being fair when people cannot agree about content.

Objections to human cloning based on the rightness of the act are of two kinds: those that hold that the act is intrinsically wrong and those that claim that the act is wrong because of potential bad consequences. Of the former, the typical objection appeals to God's will or to the naturalness of human, sexual reproduction. The second, consequentialist, objection appeals to bad consequences from human cloning to human society, to the family, or to the child originated by cloning.

Consequentialist objections concerning the welfare of the child are the most weighty moral objections to cloning and in turn are of two types: those concerning the safety of the process (physical, genetic harm to the child), and those concerning possible psychological harm to the child from either unrealistic parental expectations or a confused identity. If the safety objection is overcome one day, all the moral weight of objections to human cloning will rest on the psychological factors listed above.

If we cannot agree on the rightness of human cloning, some want to proceed to another level and discuss a process by which we can achieve good public policy. At such a point, discussions emerge about "a right to human reproduction that includes cloning" or about which government agency should oversee experimental human cloning. Those favoring the above right to human cloning want no governmental intrusion into human reproduction. Those favoring governmental oversight of human cloning include those who want regulations, such as those used by the Food and Drug Administration to supervise new drugs, and those who want state or federal control (or bans) of human cloning. Violations of the former would fall under civil law, whereas violations of the latter kind would likely be violations of criminal law.

ABOUT THE ESSAYS

The essays in this volume offer various points of view on whether we should be excited, disturbed, or indifferent to the prospect of human cloning. The essays represent both sides of this emotional issue.

James Watson's "Moving toward the Clonal Man" produced a lot of worry when it came out in the *Atlantic Monthly* magazine in 1971. It is a useful piece to remind us not to overreact to new developments in reproductive medicine.

George Johnson's "Don't Worry: A Brain Still Can't Be Cloned" is a response to some of the wilder speculations about cloning. For example, in the movie *Alien: Resurrection*, not only is Ripley's body reproduced but also her brain (even with her former memories!). Johnson, a science writer for the *New York Times*, explains why this is impossible.

In "The Wisdom of Repugnance," medical ethicist Leon Kass systematically condemns cloning, focusing especially on harms of this new form of reproduction to human society, to society, and to the identity of the child. In this vein, Kass repeats many of the objections to in vitro fertilization that he made twenty years ago in "Making Babies Revisited."

Lutheran theologian Gilbert Meilaender follows Kass's criticisms by emphasizing how human cloning would destroy the symbolic meaning of human, sexual reproduction. Meilaender traces the roots of his criticism back to the Creation story in Genesis.

Within a few days of the announcement of the cloning of Dolly, President William Clinton asked a little-known, fifteen-member bioethics commission to investigate the ethics of human cloning. The original mandate of the National Bioethics Advisory Commission (NBAC) was to study genetic privacy and the ethics of medical research. NBAC was also asked to hold hearings about human cloning so that various groups could testify, and given just ninety days to issue its conclusions. Other countries around the world already had laws making human cloning illegal, and by the time the NBAC's report was issued in June 1997, most European countries had banned human cloning. The selection from the NBAC *Report* reprinted in this volume explains the ethical objections to human cloning that various witnesses gave in hearings before the NBAC. In its conclusions (also reprinted here), NBAC explains why it concluded that a federal law should be passed to ban human cloning.

Philip Kitcher, a philosopher of science, agrees with NBAC's conclusions but realizes there might be some cases where human cloning would seem justified. However, he condemns the general option of giving parents the right to choose the genome of their children because it would create false expectations in parents that would harm the children.

George Annas, a law professor who testified before NBAC along with Meilaender, passionately condemns not only human cloning but also the whole field of unregulated assisted reproduction in medicine. Annas proposes a new federal agency, the Human Experimentation Agency, to not only review human cloning, but also to review and regulate all forms of experimental medicine.

Another law professor, John Robertson, disagrees passionately with Annas and believes that the human right to reproduce includes human cloning. Robertson's biggest target is NBAC's proposal to make human cloning a federal crime. Such a move, Robertson argues, would be without precedent in American law.

Stephen Jay Gould, the noted Harvard paleontologist and defender of evolution, doubts whether we should regard Dolly's originating cell as an "adult" cell in the ordinary sense of the term. He thinks that mammary stem cells may be more undifferentiated than other adult cells. Even if they are not, Gould thinks that human cloning raises no great new issues because we already know that conjoined twins differ in personalities and achievements. Gould also cites the work of his friend Frank Sulloway on birth order and achievement to argue that environment places a great role in the formation of the identity of a human.

Axel Kahn predicts that some medical conditions might pave the way for human cloning and argues that many innovations in the recent past have been accepted with little concern for the safety of resulting children. Kahn also shows how two parents could both have some genetic connection to a cloned child.

The objection that human cloning will be harmful, Gregory E. Pence argues, is often not justified. His questions are, "Harmful compared to what?" and "Harm to whom?" He argues that embryos cannot be harmed, and that since the high rates of embryo loss in normal reproduction and in vitro fertilization are considered acceptable, no higher standard should be set for human cloning.

R. C. Lewontin agrees that most objections to human cloning are confused. Lewontin attacks the NBAC report as being internally contradictory, sometimes seemingly endorsing genetic determinism or genetic reductionism and at other times attacking these fallacies.

In a different approach, philosopher Timothy F. Murphy examines the reasons why many gay men and lesbians have embraced human cloning. Murphy argues that many of their hopes are mistaken and stem from the prejudicial way that gay men and lesbians are treated by medicine and by the rest of society, evils which, if removed, would eliminate the desire of such gay men and lesbians for human cloning.

————————

1. Gina Kolata, "Researcher Clones Embryos of Human in Fertility Effort," *New York Times*, 26 October 1993, A1.

2. Ian Wilmut, Conference on Mammalian Cloning: Implications for Science and Society, 26 June 1997, Crystal City Marriott, Crystal City, Virginia.

3. C. Craig Venter, "Placing Cloning in the Context of the Genome," Conference on Mammalian Cloning: Implications for Science and Society, 26 June 1997, Crystal City Marriott, Crystal City, Virginia.

4. Gina Kolata, "Lab Yields Lamb with Human Gene," *New York Times,* 25 July 1997, A6.

5. Gina Kolata, "Lab Yields . . . ."

6. *The Sciences*, May/June 1997, 10.

7. Michael Specter and Gina Kolata, "After Decades and Many Missteps, Cloning Success," *New York Times*, 3 March 1997.

MOVING TOWARD THE

CLONAL MAN:

IS THIS WHAT WE WANT?

JAMES D. WATSON

This article is included here to give an historical perspective on human cloning. In the early 1970s, cloning was discussed extensively by many in medicine and bioethics. This article conveys the highly charged atmosphere that then greeted the idea of human cloning. In 1953, James Watson and Francis Crick published in Nature *their description of the double helix structure of deoxyribonucleic acid (DNA), the nucleic acid responsible for transmitting hereditary characteristics. At the time, Watson was only twenty-five years old. This description included the basic mechanism for copying genetic material from one cell to another and, in reproduction, from one generation to another. Watson and Crick's spectacular discovery moved the study of genetics from observational inference to molecular biology, and the duo received (with another collaborator, Maurice Wilkins) the Nobel Prize for medicine in 1962.*

Watson served in the Biology Department at Harvard University from 1955 to 1976, after which he was director of the famous Cold Spring Harbor Laboratory on Long Island. This lab carried out ground-breaking high-level research on cancer, genetics, molecular biology, cellular biology, and neurology. From 1989 to 1992, he was director of the National Center for Human Genome Research of the National Institutes of Health, having been a key player in convincing Congress to fund the Human Genome Project. He later quit this project amid a controversy over patent rights.

Watson is one of the most famous people in science, but when he moved from writing about science to writing about the ethical implications of science, his writings became controversial. In 1971, in the article reprinted here, Wat-

son predicted that dangerous events would occur if physicians tried to use in vitro fertilization to originate a human baby. More important to us, he also seemed to confuse in vitro fertilization with originating a human by cloning. In his conclusion here, he urged a world-wide ban on research that might lead to originating a human by cloning.

THE NOTION THAT MAN MIGHT SOMETIME SOON BE REPRODUCED asexually upsets many people. The main public effect of the remarkable clonal frog produced some ten years ago in Oxford by the zoologist John Gurdon has not been awe of the elegant scientific implication of this frog's existence, but fear that a similar experiment might someday be done with human cells. Until recently, however, this foreboding has seemed more like a science fiction scenario than a real problem which the human race has to live with.

For the embryological development of man does not occur free in the placid environment of a freshwater pond, in which a frog's eggs normally turn into tadpoles and then into mature frogs. Instead, the crucial steps in human embryology always occur in the highly inaccessible womb of a human female. There the growing fetus enlarges unseen, and effectively out of range of almost any manipulation except that which is deliberately designed to abort its existence. As long as all humans develop in this manner, there is no way to take the various steps necessary to insert an adult diploid nucleus from a pre-existing human into a human egg whose maternal genetic material has previously been removed. Given the continuation of the normal processes of conception and development, the idea that we might have a world populated by people whose genetic material was identical to that of previously existing people can belong only to the domain of the novelist or moviemaker, not to that of pragmatic scientists who must think only about things which can happen.

Today, however, we must face up to the fact that the unexpectedly rapid progress of R. G. Edwards and P. S. Steptoe in working out the conditions for routine test-tube conception of human eggs means that human embryological development need no longer be a process shrouded in secrecy. It can become instead an event wide-open to a variety of experimental manipulations. Already the two scientists have developed many embryos to the eight-cell stage, and a few more into blastocysts, the stage where successful implantation into a human uterus should not be too difficult to achieve. In fact, Edwards and Steptoe hope to accomplish implantation and subsequent growth into a normal baby within the coming year.

The question naturally arises, why should any woman willingly submit to the laparoscopy operation which yields the eggs to be used in test-tube conceptions? There is clearly some danger involved every time Steptoe operates. Nonetheless, he and Edwards believe that the risks are more than counterbalanced by the fact that their research may develop methods which could make their patients able to bear children. All their patients, though having normal menstrual cycles, are infertile, many because they have blocked oviducts which prevent passage of eggs into the uterus. If so, *in vitro* growth of their eggs up to the blastocyst stage may circumvent infertility, thereby allowing normal childbirth. Moreover, since the sex of a blastocyst is easily determined by chromosomal analysis, such women would have the possibility of deciding whether to give birth to a boy or a girl.

Clearly, if Edwards and Steptoe succeed, their success will be followed up in many other places. The number of such infertile women, while small on a relative percentage basis, is likely to be large on an absolute basis. Within the United States there could be 100,000 or so women who would like a similar chance to have their own babies. At the same time, we must anticipate strong, if not hysterical, reactions from many quarters. The certainty that the ready availability of this medical technique will open up the possibility of hiring out unrelated women to carry a given baby to term is bound to outrage many people. For there is absolutely no reason why the blastocyst need be implanted in the same woman from whom the pre-ovulatory eggs were obtained. Many women with anatomical complications which prohibit successful childbearing might be strongly tempted to find a suitable surrogate. And it is easy to imagine that other women who just don't want the discomforts of pregnancy would also seek this very different form of motherhood. Of even greater concern would be the potentialities for misuse by an inhumane totalitarian government.

Some very hard decisions may soon be upon us. It is not obvious, for example, that the vague potential of abhorrent misuse should weigh more strongly than the unhappiness which thousands of married couples feel when they are unable to have their own children. Different societies are likely to view the matter differently, and it would be surprising if all should come to the same conclusion. We must, therefore, assume that techniques for the *in vitro* manipulation of human eggs are likely to become general medical practice, capable of routine performance in many major countries, within some ten to twenty years.

The situation would then be ripe for extensive efforts, either legal or illegal, at human cloning. But for such experiments to be successful, techniques would have to be developed which allow the insertion of adult diploid nuclei into human eggs which previously have had their maternal haploid nucleus removed. At first sight, this task is a very tall order since human eggs are much smaller than those of frogs, the only vertebrates which have so far been cloned. Insertion by micropipettes, the device used in the case of the frog, is always likely to damage human eggs irreversibly. Recently, however, the development of simple techniques for fusing animal cells has raised the strong possibility that further refinements of the cell-fusion method will allow the routine introduction of human diploid nuclei into enucleated human eggs. Activation of such eggs to divide to become blastocysts, followed by implantation into suitable uteri, should lead to the development of healthy fetuses, and subsequent normal-appearing babies.

The growing up to adulthood of these first clonal humans could be a very startling event, a fact already appreciated by many magazine editors, one of whom commissioned a cover with multiple copies of Ringo Starr, another of whom gave us overblown multiple likenesses of the current sex goddess, Raquel Welch. It takes little imagination to perceive that different people will have highly different fantasies, some perhaps imagining the existence of countless people with the features of Picasso or Frank Sinatra or Walt Frazier or Doris Day. And would monarchs like the Shah of Iran, knowing they might never be able to have a normal male heir, consider the possibility of having a son whose genetic constitution would be identical to their own?

Clearly, even more bizarre possibilities can be thought of, and so we might have expected that many biologists, particularly those whose work impinges upon this possibility, would seriously ponder its implication, and begin a dialogue which would educate the world's citizens and offer suggestions which our legislative bodies might consider in framing national science policies. On the whole, however, this has not happened. Though a number of scientific papers devoted to the problem of genetic engineering have casually mentioned that clonal reproduction may someday be with us, the discussion to which I am party has been so vague and devoid of meaningful time estimates as to be virtually soporific.

Does this effective silence imply a conspiracy to keep the general public unaware of a potential threat to their basic ways of life? Could it be motivated by fear that the general reaction will be a further damning of all science, thereby decreasing even more the limited money available for pure research? Or does it merely tell us that most scientists do live such an ivory-tower existence that they are capable of thinking rationally only about pure science, dismissing more practical matters as subjects for the lawyers, students, clergy, and politicians to face up to?

One or both of these possibilities may explain why more scientists have not taken cloning before the public. The main reason, I suspect, is that the prospect to most biologists still looks too remote and chancy—not worthy of immediate attention when other matters, like nuclear-weapon overproliferation and pesticide and auto-exhaust pollution, present society with immediate threats to its orderly continuation. Though scientists as a group form the most future-oriented of all professions, there are few of us who concentrate on events un-likely to become reality within the next decade or two.

To almost all the intellectually most adventurous geneticists, the seemingly distant time when cloning might first occur is more to the point than its far-reaching implication, were it to be practiced seriously. For example, Stanford's celebrated geneticist, Joshua Lederberg, among the first to talk about cloning as a practical matter, now seems bored with further talk, implying that we should channel our limited influence as public citizens to the prevention of the wide-scale, irreversible damage to our genetic material that is now occurring through increasing exposure to man-created mutagenic compounds. To him, serious talk about cloning is essentially crying wolf when a tiger is already inside the walls.

This position, however, fails to allow for what I believe will be a frenetic rush to do experimental manipulation with human eggs once they have become a readily available commodity. And that is what they will be within several years after Edwards-Steptoe methods lead to the birth of the first healthy baby by a previously infertile woman. Isolated human eggs will be found in hundreds of hospitals, and given the fact that Steptoe's laparoscopy technique frequently yields several eggs from a single woman donor, not all of the eggs so obtained, even if they could be cultured to the blastocyst stage, would ever be reim-planted into female bodies. Most of these excess eggs would likely be used for a variety of valid experimental purposes, many, for example, to perfect the Edwards-Steptoe techniques. Others could be devoted to finding methods for curing certain genetic diseases, conceivably through use of cell-fusion methods which now seem to be the correct route to cloning. The temptation to try cloning itself thus will always be close at hand.

No reason, of course, dictates that such cloning experiments need occur. Most of the medical people capable of such experimentation would probably steer clear of any step which looked as though its real purpose were to clone. But it would be shortsighted to assume that everyone would instinctively recoil from such purposes. Some people may sincerely believe the world desperately needs many copies of really exceptional people if we are to fight our way out of the ever-increasing computer-mediated complexity that makes our individual brains so frequently inadequate.

Moreover, given the widespread development of the safe clinical procedures for handling human eggs, cloning experiments would not be prohibitively ex-pensive. They need not be restricted to the superpowers. All smaller countries now possess the resources required for eventual success. Furthermore, there need not exist the coercion of a totalitarian state to provide the surrogate moth-ers. There already are such widespread divergences regarding the sacredness of the act of human reproduction that the boring meaninglessness of the lives of many women would be sufficient cause for their willingness to participate in such experimentation, be it legal or illegal. Thus, if the matter proceeds in its current nondirected fashion, a human being born of clonal reproduction most likely will appear on the earth within the next twenty to fifty years, and even sooner, if some nation should actively promote the venture.

The first reaction of most people to the arrival of these asexually produced children, I suspect, would be one of despair. The nature of the bond between parents and their children, not to mention everyone's values about the individual's uniqueness, could be changed beyond recognition, and by a science which they never understood but which until recently appeared to provide more good than harm. Certainly to many people, particularly those with strong religious backgrounds, our most sensible course of action would be to de-emphasize all those forms of research which would circumvent the normal sexual reproductive process. If this step were taken, experiments on cell fusion might no longer be supported by federal funds or tax-exempt organizations. Prohibition of such research would most certainly put off the day when diploid nuclei could satisfactorily be inserted into enucleated human eggs. Even more effective would be to take steps quickly to make illegal, or to reaffirm the illegality of, any experimental work with human embryos.

Neither of the prohibitions, however, is likely to take place. In the first place, the cell-fusion technique now offers one of the best avenues for understanding the genetic basis of cancer. Today, all over the world, cancer cells are being fused with normal cells to pinpoint those specific chromosomes responsible for given forms of cancer. In addition, fusion techniques are the basis of many genetic efforts to unravel the biochemistry of diseases like cystic fibrosis or multiple sclerosis. Any attempts now to stop such work using the argument that cloning represents a greater threat than a disease like cancer is likely to be considered irresponsible by virtually anyone able to understand the matter.

Though more people would initially go along with a prohibition of work on human embryos, many may have a change of heart when they ponder the mess which the population explosion poses. The current projections are so horrendous that responsible people are likely to consider the need for more basic embryological facts much more relevant to our self-interest than the not-very-immediate threat of a few clonal men existing some decades ahead. And the potentially militant lobby of infertile couples who see test-tube conception as their only route to the joys of raising children of their own making would carry even more weight. So, scientists like Edwards are likely to get a go-ahead signal even if, almost perversely, the immediate consequences of their "population-money"-supported research will be the production of still more babies.

Complicating any effort at effective legislative guidance is the multiplicity of places where work like Edwards' could occur, thereby making unlikely the possibility that such manipulations would have the same legal (or illegal) status throughout the world. We must assume that if Edwards and Steptoe produce a really workable method for restoring fertility, large numbers of women will search out those places where it is legal (or possible), just as now they search out places where abortions can be easily obtained.

Thus, all nations formulating policies to handle the implications of *in vitro* human embryo experimentation must realize that the problem is essentially an international one. Even if one or more countries should stop such research, their action could effectively be neutralized by the response of a neighboring country. This most disconcerting impotence also holds for the United States. If our congressional representatives, upon learning where the matter now stands, should decide that they want none of it and pass very strict laws against human embryo experimentation, their action would not seriously set back the current scientific and medical momentum which brings us close to the possibility of surrogate mothers, if not human clonal reproduction. This is because the relevant experiments are being done not in the United States, but largely in England. That is partly a matter of chance, but also a consequence of the advanced state of English cell biology, which in certain areas is far more adventurous and imaginative than its American counterpart. There is no American university which has the strength in experimental embryology that Oxford possesses.

We must not assume, however, that today the important decisions lie only before the British government. Very soon we must anticipate that a number of biologists and clinicians of other countries, sensing the potential excitement, will move into this area of science. So even if the current English effort were stifled, similar experimentation could soon begin elsewhere. Thus it appears to me most desirable that as many people as possible be informed about the new ways of human reproduction and their potential consequences, both good and bad.

This is a matter far too important to be left solely in the hands of the scientific and medical communities. The belief that surrogate mothers and clonal babies are inevitable because science always moves forward, an attitude expressed to me recently by a scientific colleague, represents a form of laissez-faire nonsense dismally reminiscent of the creed that American business, if left to itself, will solve everybody's problems. Just as the success of a corporate body in making money need not set the human condition ahead, neither does every scientific advance automatically make our lives more "meaningful." No doubt the person whose experimental skill will eventually bring forth a clonal baby will be given wide notoriety. But the child who grows up knowing that the world wants another Picasso may view his creator in a different light.

I would thus hope that over the next decade wide-reaching discussion would occur, at the informal as well as formal legislative level, about the manifold problems which are bound to arise if test-tube conception becomes a common occurrence. A blanket declaration of the worldwide illegality of human cloning might be one result of a serious effort to ask the world in which direction it wished to move. Admittedly the vast effort, required for even the most limited international arrangement, will turn off some people—those who believe the matter is of marginal importance now, and that it is a red herring designed to take our minds off our callous attitudes toward war, poverty, and racial prejudice. But if we do not think about it now, the possibility of our having a free choice will one day suddenly be gone.

DON'T WORRY:

A BRAIN STILL

CAN'T BE CLONED

GEORGE JOHNSON

George Johnson is a veteran science writer for the New York Times. *He also moderated a forum on the ethics of human cloning for the* New York Times' *discussion group on the Internet. In that discussion, one fan praised the article reprinted below as one of the finest examples of writing about science that he had read on cloning.*

EXPLORERS RETURNING FROM DISTANT LANDS TELL OF ABORIGINES so afraid of cameras that they recoil from the sight of a lens as if they were looking down the barrel of a gun. Taking their picture, they fear, is the same as stealing their soul.

You might as well just shoot them dead on the spot. Knowing that a photograph is only skin deep, people in the developed lands find such terror absurd. But the fear that one's very identity might be stolen, that one could cease to be an individual, runs deep even in places where cameras seem benign.

The queasiness many people feel over the news last week that a scientist in Scotland has made a carbon copy of a sheep comes down to this: if a cell can be taken from a human being and used to create a genetically identical double, then any of us could lose our uniqueness. One would no longer be a self.

There are plenty of other reasons to worry about this new divide the biologists have trampled across. Nightmare of the week goes to those who imagine docile flocks of enslaved clones raised for body parts.

But the most fundamental fear is that the soul will be taken by this penetrating new photography called cloning. And here, at least, the notion is just as superstitious as the aborigines'. There is one part of life biotechnology will never touch. While it is possible to clone a body, it is impossible to clone a brain.

That each creature from microbe to man is unique in all the world is amazing when you consider that every life form is assembled from the same identical building blocks. Every electron in the universe is indistinguishable, by definition.

You can't tell one from the other by examining it for nicks and scratches. All protons and all neutrons are also precisely the same.

And when you put these three kinds of particles together to make atoms, there is still no individuality. Every carbon atom and every hydrogen atom is the same. When atoms are strung together into complex molecules—the enzymes and other proteins—this uniformity begins to break down. Minor variations occur.

But it is at the next step up the ladder that something strange and wonderful happens. There are so many ways molecules can be combined into the complex little machines called cells that no two of them can be exactly alike.

Even cloned cells, with identical sets of genes, vary somewhat in shape or coloration. The variations are so subtle they can usually be ignored. But when cells are combined to form organisms, the differences become overwhelming. A threshold is crossed and individuality is born.

Two genetically identical twins inside a womb will unfold in slightly different ways. The shape of the kidneys or the curve of the skull won't be quite the same. The differences are small enough that an organ from one twin can probably be transplanted into the other. But with the organs called brains the differences become profound.

All a body's tissues—bone, skin, muscle, and so forth—are made by taking the same kind of cell and repeating it over and over again. But with brain tissue there is no such monotony.

The precise layout of the cells, which neuron is connected to which, makes all the difference. Linked one with the other, through the junctions called synapses, neurons form the whorls of circuitry whose twists and turns make us who we are.

In the reigning metaphor, the genome, the coils of DNA that carry the genetic information, can be thought of as a computer directing the assembly of the embryo. Back-of-the-envelope calculations show how much information a human genome contains and how much information is required to specify the trillions of connections in a single brain.

The conclusion is inescapable: the problem of wiring up a brain is so complex that it is beyond the power of the genomic computer.

The best the genes can do is indicate the rough layout of the wiring, the general shape of the brain. Neurons, in this early stage, are thrown together more or less at random and then left to their own devices.

After birth, experience makes and breaks connections, pruning the thicket into precise circuitry. From the very beginning, what's in the genes is different from what's in the brain. And the gulf continues to widen as the brain matures.

The genes still exert their influence—some of the brain's circuitry is hard-wired from the start and immutable. People don't have to learn to want food or sex. But as the new connections form, the mind floating higher and higher above the genetic machinery like a helium balloon, people learn to circumvent the baser instincts in individual ways.

Even genetically identical twins, natural clones, are born with different neural tangles. Subtle variations in the way the connections were originally slapped together might make one twin particularly fascinated by twinkling lights, the other drawn to certain patterns of sounds.

Even if the twins were kept in the same room for days, these natural predilections would drive them each in different directions. Experience, pouring in through the senses, would cause unique circuitry to form. Once the twins left the room, the differences between them would increase.

Send one twin around the block clockwise and the other counterclockwise and they would return with more divergent brains. For artificial clones the variations would accumulate even faster, for they would be born years apart, into different worlds.

Photography is only skin deep. Cloning is only gene deep. But what about the ultimate cloning—copying synapse by synapse a human brain?

If such a technological feat were ever possible, for one brief instant we might have two identical minds. But then suppose neuron No. 20478288 were to fire randomly in brain 1 and not in brain 2. The tiny spasm would set off a cascade that reshaped some circuitry, and there would be two individuals again.

We each carry in our heads complexity beyond imagining and beyond duplication. Even a hard-core materialist might agree that, in that sense, everyone has a soul.

THE WISDOM

OF REPUGNANCE

LEON KASS

In 1967, a young molecular biologist named Leon Kass wrote a letter to the Washington Post, protesting a previous column by Nobel laureate Joshua Lederberg that defended human cloning along eugenic lines. Ever since, that biologist has been writing a lot about bioethics. So it is no surprise that he is still attacking human cloning. Indeed, the article by Leon Kass reprinted here is the most sustained and systematic attack to date on the permissibility of cloning.

In the early 1970s, Kass was one of the foremost opponents of in vitro fertilization, arguing that this form of reproduction might harm the child, the family, and society. In the following piece Kass uses premises that can be accepted by both religious and secular people. His defense of human sexual reproduction rests on its naturalness, its place in our traditional social systems, and its place in evolution. Kass's most sustained criticism is that a child originated by cloning would be saddled with unrealistic, hurtful expectations by adults and suffer a confused lineage and identity.

Leon Kass is the Addie Clark Harding Professor in The College and the Committee on Social Thought at the University of Chicago. He has written the classic articles in bioethics, "Making Babies" and "Making Babies Revisited," as well as Toward a More Natural Science: Biology and Human Affairs.

OUR HABIT OF DELIGHTING IN NEWS OF SCIENTIFIC AND TECHNO-logical breakthroughs has been sorely challenged by the birth announcement of a sheep named Dolly. Though Dolly shares with previous sheep the "softest clothing, woolly, bright," William Blake's question, "Little Lamb, who made thee?" has for her a radically different answer: Dolly was, quite literally, made. She is the work not of nature or nature's God but of man, an Englishman, Ian

Wilmut, and his fellow scientists. What's more, Dolly came into being not only asexually—ironically, just like "He [who] calls Himself a Lamb"—but also as the genetically identical copy (and the perfect incarnation of the form or blueprint) of a mature ewe, of whom she is a clone. This long-awaited yet not quite expected success in cloning a mammal raised immediately the prospect—and the specter—of cloning human beings: "I a child and Thou a lamb," despite our differences, have always been equal candidates for creative making, only now, by means of cloning, we may both spring from the hand of man playing at being God.

After an initial flurry of expert comment and public consternation, with opinion polls showing overwhelming opposition to cloning human beings, President Clinton ordered a ban on all federal support for human cloning research (even though none was being supported) and charged the National Bioethics Advisory Commission to report in ninety days on the ethics of human cloning research. The commission (an eighteen-member panel, evenly balanced between scientists and nonscientists, appointed by the president and reporting to the National Science and Technology Council) invited testimony from scientists, religious thinkers and bioethicists, as well as from the general public. It is now deliberating about what it should recommend, both as a matter of ethics and as a matter of public policy.

Congress is awaiting the commission's report, and is poised to act. Bills to prohibit the use of federal funds for human cloning research have been introduced in the House of Representatives and the Senate; and another bill, in the House, would make it illegal "for any person to use a human somatic cell for the process of producing a human clone." A fateful decision is at hand. To clone or not to clone a human being is no longer an academic question.

TAKING CLONING SERIOUSLY, THEN AND NOW

Cloning first came to public attention roughly thirty years ago, following the successful asexual production, in England, of a clutch of tadpole clones by the technique of nuclear transplantation. The individual largely responsible for bringing the prospect and promise of human cloning to public notice was Joshua Lederberg, a Nobel Laureate geneticist and a man of large vision. In 1966, Lederberg wrote a remarkable article in *The American Naturalist* detailing the eugenic advantages of human cloning and other forms of genetic engineering, and the following year he devoted a column in *The Washington Post*, where he wrote regularly on science and society, to the prospect of human cloning. He suggested that cloning could help us overcome the unpredictable variety that still rules human reproduction, and allow us to benefit from perpetuating superior genetic endowments. These writings sparked a small public debate in which I became a participant. At the time a young researcher in molecular biology at the National Institutes of Health (NIH), I wrote a reply to the *Post*, arguing against Lederberg's amoral treatment of this morally weighty subject and insisting on the urgency of confronting a series of questions and objections, culminating in the suggestion that "the programmed reproduction of man will, in fact, dehumanize him."

Much has happened in the intervening years. It has become harder, not easier, to discern the true meaning of human cloning. We have in some sense been softened up to the idea—through movies, cartoons, jokes and intermittent commentary in the mass media, some serious, most lighthearted. We have become accustomed to new practices in human reproduction: not just in vitro fertilization, but also embryo manipulation, embryo donation and surrogate pregnancy. Animal biotechnology has yielded transgenic animals and a burgeoning science of genetic engineering, easily and soon to be transferable to humans.

Even more important, changes in the broader culture make it now vastly more difficult to express a common and respectful understanding of sexuality, procreation, nascent life, family, and the meaning of motherhood, fatherhood and the links between the generations. Twenty-five years ago, abortion was still largely illegal and thought to be immoral, the sexual revolution (made possible by the extramarital use of the pill) was still in its infancy, and few had yet heard about the reproductive rights of single women, homosexual men and lesbians. (Never mind shameless memoirs about one's own incest!) Then one could argue, without embarrassment, that the new technologies of human reproduction—babies without sex—and their confounding of normal kin relations—who's the mother: the egg donor, the surrogate who carries and delivers, or the one who rears?—would "undermine the justification and support that biological parenthood gives to the monogamous marriage." Today, defenders of stable, monogamous marriage risk charges of giving offense to those adults who are living in "new family forms" or to those children who, even without the benefit of assisted reproduction, have acquired either three or four parents or one or none at all. Today, one must even apologize for voicing opinions that twenty-five years ago were nearly universally regarded as the core of our culture's wisdom on these matters. In a world whose once-given natural boundaries are blurred by technological change and whose moral boundaries are seemingly up for grabs, it is much more difficult to make persuasive the still compelling case against cloning human beings. As Raskolnikov put it, "man gets used to everything—the beast!"

Indeed, perhaps the most depressing feature of the discussions that immediately followed the news about Dolly was their ironical tone, their genial cynicism, their moral fatigue: "AN UDDER WAY OF MAKING LAMBS" (*Nature*), "WHO WILL CASH IN ON BREAKTHROUGH IN CLONING?" (*The Wall Street Journal*), "IS CLONING BAAAAAAAAD?" (*The Chicago Tribune*). Gone from the scene are the wise and courageous voices of Theodosius Dobzhansky (genetics), Hans Jonas (philosophy) and Paul Ramsey (theology) who, only twenty-five years ago, all made powerful moral arguments against ever cloning a human being. We are now too sophisticated for such argumentation; we wouldn't be caught in public with a strong moral stance, never mind an absolutist one. We are all, or almost all, postmodernists now.

Cloning turns out to be the perfect embodiment of the ruling opinions of our new age. Thanks to the sexual revolution, we are able to deny in practice, and increasingly in thought, the inherent procreative teleology of sexuality itself. But, if sex has no intrinsic connection to generating babies, babies need have no necessary connection to sex. Thanks to feminism and the gay rights movement, we are increasingly encouraged to treat the natural heterosexual difference and its preeminence as a matter of "cultural construction." But if male and female are not normatively complementary and generatively significant, babies need not come from male and female complementarity. Thanks to the prominence and the acceptability of divorce and out-of-wedlock births, stable, monogamous marriage as the ideal home for procreation is no longer the agreed-upon cultural norm. For this new dispensation, the clone is the ideal emblem: the ultimate "single-parent child."

Thanks to our belief that all children should be *wanted* children (the more high-minded principle we use to justify contraception and abortion), sooner or later only those children who fulfill our wants will be fully acceptable. Through cloning, we can work our wants and wills on the very identity of our children, exercising control as never before. Thanks to modern notions of individualism and the rate of cultural change, we see ourselves not as linked to ancestors and defined by traditions, but as projects for our own self-creation, not only as self-made men but also man-made selves; and self-cloning is simply an extension of such rootless and narcissistic self-re-creation.

Unwilling to acknowledge our debt to the past and unwilling to embrace the uncertainties and the limitations of the future, we have a false relation to both: cloning personifies our desire fully to control the future, while being subject to no controls ourselves. Enchanted and enslaved by the glamour of technology, we have lost our awe and wonder before the deep mysteries of nature and of life. We cheerfully take our own beginnings in our hands and, like the last man, we blink.

Part of the blame for our complacency lies, sadly, with the field of bioethics itself, and its claim to expertise in these moral matters. Bioethics was founded by people who understood that the new biology touched and threatened the deepest matters of our humanity: bodily integrity, identity and individuality, lineage and kinship, freedom and self-command, eros and aspiration, and the relations and strivings of body and soul. With its capture by analytic philosophy, however, and its inevitable routinization and professionalization, the field has by and large come to content itself with analyzing moral arguments, reacting to new technological developments and taking on emerging issues of public policy, all performed with a naïve faith that the evils we fear can all be avoided by compassion, regulation and a respect for autonomy. Bioethics has made some major contributions in the protection of human subjects and in other areas where personal freedom is threatened; but its practitioners, with few exceptions, have turned the big human questions into pretty thin gruel.

One reason for this is that the piecemeal formation of public policy tends to grind down large questions of morals into small questions of procedure. Many of the country's leading bioethicists have served on national commissions or state task forces and advisory boards, where understandably, they have found utilitarianism to be the only ethical vocabulary acceptable to all participants in discussing issues of law, regulation and public policy. As many of these commissions have been either officially under the aegis of NIH or the Health and Human Services Department, or otherwise dominated by powerful voices for scientific progress, the ethicists have for the most part been content, after some "values clarification" and wringing of hands, to pronounce their blessings upon the inevitable. Indeed, it is the bioethicists, not the scientists, who are now the most articulate defenders of human cloning: the two witnesses testifying before the National Bioethics Advisory Commission in favor of cloning human beings were bioethicists, eager to rebut what they regard as the irrational concerns of those of us in opposition. One wonders whether this commission, constituted like the previous commissions, can tear itself sufficiently free from the accommodationist pattern of rubber-stamping all technical innovation, in the mistaken belief that all other goods must bow down before the gods of better health and scientific advance.

If it is to do so, the commission must first persuade itself, as we all should persuade ourselves, not to be complacent about what is at issue here. Human cloning, though it is in some respects continuous with previous reproductive technologies, also represents something radically new, in itself and in its easily foreseeable consequences. The stakes are very high indeed. I exaggerate, but in the direction of the truth, when I insist that we are faced with having to decide nothing less than whether human procreation is going to remain human, whether children are going to be made rather than begotten, whether it is a good thing, humanly speaking, to say yes in principle to the road which leads (at best) to the dehumanized rationality of *Brave New World*. This is not business as usual, to be fretted about for a while but finally to be given our seal of approval. We must rise to the occasion and make our judgments as if the future of our humanity hangs in the balance. For so it does.

THE STATE OF THE ART

If we should not underestimate the significance of human cloning, neither should we exaggerate its imminence or misunderstand just what is involved. The procedure is conceptually simple. The nucleus of a mature but unfertilized egg is removed and replaced with a nucleus obtained from a specialized cell of an adult (or fetal) organism (in Dolly's case, the donor nucleus came from mammary gland epithelium). Since almost all the hereditary material of a cell is contained within its nucleus, the renucleated egg and the individual into which this egg develops are genetically identical to the organism that was the source of the transferred nucleus. An unlimited number of genetically identical individuals—clones—could be produced by nuclear transfer. In principle, any person, male or female, newborn or adult, could be cloned, and in any quantity. With laboratory cultivation and storage of tissues, cells outliving their sources make it possible even to clone the dead.

The technical stumbling block, overcome by Wilmut and his colleagues, was to find a means of reprogramming the state of the DNA in the donor cells, reversing its differentiated expression and restoring its full totipotency, so that it could again direct the entire process of producing a mature organism. Now that this problem has been solved, we should expect a rush to develop cloning for other animals, especially livestock, in order to propagate in perpetuity the champion meat or milk producers. Though exactly how soon someone will succeed in cloning a human being is anybody's guess, Wilmut's technique, almost certainly applicable to humans, makes *attempting* the feat an imminent possibility.

Yet some cautions are in order and some possible misconceptions need correcting. For a start, cloning is not Xeroxing. As has been reassuringly reiterated, the clone of Mel Gibson, though his genetic double, would enter the world hairless, toothless and peeing in his diapers, just like any other human infant. Moreover, the success rate, at least at first, will probably not be very high: the British transferred 277 adult nuclei into enucleated sheep eggs, and implanted twenty-nine clonal embryos, but they achieved the birth of only one live lamb clone. For this reason, among others, it is unlikely that, at least for now, the practice would be very popular, and there is no immediate worry of mass-scale production of multicopies. The need of repeated surgery to obtain eggs and, more crucially, of numerous borrowed wombs for implantation will surely limit use, as will the expense; besides, almost everyone who is able will doubtless prefer nature's sexier way of conceiving.

Still, for the tens of thousands of people already sustaining over 200 assisted reproduction clinics in the United States and already availing themselves of in vitro fertilization, intracytoplasmic sperm injection and other techniques of assisted reproduction, cloning would be an option with virtually no added fuss (especially when the success rate improves). Should commercial interests develop in "nucleus-banking," as they have in sperm-banking; should famous athletes or other celebrities decide to market their DNA the way they now market their autographs and just about everything else; should techniques of embryo and germline genetic testing and manipulation arrive as anticipated, increasing the use of laboratory assistance in order to obtain "better" babies—should all this come to pass, then cloning, if it is permitted, could become more than a marginal practice simply on the basis of free reproductive choice, even without any social encouragement to upgrade the gene pool or to replicate superior types. Moreover, if laboratory research on human cloning proceeds, even without any intention to produce cloned humans, the existence of cloned human embryos in the laboratory, created to begin with only for research purposes, would surely pave the way for later baby-making implantations.

In anticipation of human cloning, apologists and proponents have already made clear possible uses of the perfected technology, ranging from the sentimental and compassionate to the grandiose. They include: providing a child for an infertile couple; "replacing" a beloved spouse or child who is dying or has died; avoiding the risk of genetic disease; permitting reproduction for homosexual men and lesbians who want nothing sexual to do with the opposite sex; securing a genetically identical source of organs or tissues perfectly suitable for transplantation; getting a child with a genotype of one's own choosing, not excluding oneself; replicating individuals of great genius, talent or beauty—having a child who really could "be like Mike"; and creating large sets of genetically identical humans suitable for research on, for instance, the question of nature versus nurture, or for special missions in peace and war (not excluding espionage), in which using identical humans would be an advantage. Most people who envision the cloning of human beings, of course, want none of these scenarios. That they cannot say why is not surprising. What is surprising, and welcome, is that, in our cynical age, they are saying anything at all.

THE WISDOM OF REPUGNANCE

"Offensive." "Grotesque." "Revolting." "Repugnant." "Repulsive." These are the words most commonly heard regarding the prospect of human cloning. Such reactions come both from the man or woman in the street and from the intellectuals, from believers and atheists, from humanists and scientists. Even Dolly's creator has said he "would find it offensive" to clone a human being.

People are repelled by many aspects of human cloning. They recoil from the prospect of mass production of human beings, with large clones of look-alikes, compromised in their individuality; the idea of father-son or mother-daughter twins; the bizarre prospects of a woman giving birth to and rearing a genetic copy of herself, her spouse or even her deceased father or mother; the grotesqueness of conceiving a child as an exact replacement for another who has died; the utilitarian creation of embryonic genetic duplicates of oneself, to be frozen away or created when necessary, in case of need for homologous tissues or organs for transplantation; the narcissism of those who would clone themselves and the arrogance of others who think they know who deserves to be cloned or which genotype any child-to-be should be thrilled to receive; the Frankensteinian hubris to create human life and increasingly to control its destiny; man playing God. Almost no one finds any of the suggested reasons for human cloning compelling; almost everyone anticipates its possible misuses and abuses. Moreover, many people feel oppressed by the sense that there is probably nothing we can do to prevent it from happening. This makes the prospect all the more revolting.

Revulsion is not an argument; and some of yesterday's repugnances are today calmly accepted—though, one must add, not always for the better. In crucial cases, however, repugnance is the emotional expression of deep wisdom, be-yond reason's power fully to articulate it. Can anyone really give an argument fully adequate to the horror which is father-daughter incest (even with con-sent), or having sex with animals, or mutilating a corpse, or eating human flesh, or even just (just!) raping or murdering another human being? Would any-body's failure to give full rational justification for his or her revulsion at these practices make that revulsion ethically suspect? Not at all. On the contrary, we are suspicious of those who think that they can rationalize away our horror, say, by trying to explain the enormity of incest with arguments only about the ge-netic risks of inbreeding.

The repugnance at human cloning belongs in this category. We are repelled by the prospect of cloning human beings not because of the strangeness or novelty of the undertaking, but because we intuit and feel, immediately and without argument, the violation of things that we rightfully hold dear. Repug-nance, here as elsewhere, revolts against the excesses of human willfulness, warning us not to transgress what is unspeakably profound. Indeed, in this age in which everything is held to be permissible so long as it is freely done, in which our given human nature no longer commands respect, in which our bodies are regarded as mere instruments of our autonomous rational wills, repugnance may be the only voice left that speaks up to defend the central core of our humanity. Shallow are the souls that have forgotten how to shudder.

The goods protected by repugnance are generally overlooked by our customary ways of approaching all new biomedical technologies. The way we evaluate cloning ethically will in fact be shaped by how we characterize it descriptively, by the context into which we place it, and by the perspective from which we view it. The first task for ethics is proper description. And here is where our failure begins.

Typically, cloning is discussed in one or more of three familiar contexts, which one might call the technological, the liberal and the meliorist. Under the first, cloning will be seen as an extension of existing techniques for assisting reproduction and determining the genetic makeup of children. Like them, cloning is to be regarded as a neutral technique, with no inherent meaning or goodness, but subject to multiple uses, some good, some bad. The morality of cloning thus depends absolutely on the goodness or badness of the motives and intentions of the cloners: as one bioethicist defender of cloning puts it, "the ethics must be judged [only] by the way the parents nurture and rear their resulting child and whether they bestow the same love and affection on a child brought into existence by a technique of assisted reproduction as they would on a child born in the usual way."

The liberal (or libertarian or liberationist) perspective sets cloning in the context of rights, freedoms and personal empowerment. Cloning is just a new option for exercising an individual's right to reproduce or to have the kind of child that he or she wants. Alternatively, cloning enhances our liberation (especially women's liberation) from the confines of nature, the vagaries of chance, or the necessity for sexual mating. Indeed, it liberates women from the need for men altogether, for the process requires only eggs, nuclei and (for the time being) uteri—plus, of course, a healthy dose of our (allegedly "masculine") manipulative science that likes to do all these things to mother nature and nature's mothers. For those who hold this outlook, the only moral restraints on cloning are adequately informed consent and the avoidance of bodily harm. If no one is cloned without her consent, and if the clonant is not physically damaged, then the liberal conditions for licit, hence moral, conduct are met. Worries that go beyond violating the will or maiming the body are dismissed as "symbolic"—which is to say, unreal.

The meliorist perspective embraces valetudinarians and also eugenicists. The latter were formerly more vocal in these discussions, but they are now generally happy to see their goals advanced under the less threatening banners of freedom and technological growth. These people see in cloning a new prospect for improving human beings—minimally, by ensuring the perpetuation of healthy individuals by avoiding the risks of genetic disease inherent in the lottery of sex, and maximally, by producing "optimum babies," preserving outstanding genetic material, and (with the help of soon-to-come techniques for precise genetic engineering) enhancing inborn human capacities on many fronts. Here the morality of cloning as a means is justified solely by the excellence of the end, that is, by the outstanding traits or individuals cloned—beauty, or brawn, or brains.

These three approaches, all quintessentially American and all perfectly fine in their places, are sorely wanting as approaches to human procreation. It is, to say the least, grossly distorting to view the wondrous mysteries of birth, renewal and individuality, and the deep meaning of parent-child relations, largely through the lens of our reductive science and its potent technologies. Similarly, considering reproduction (and the intimate relations of family life!) primarily under the political-legal, adversarial and individualistic notion of rights can only undermine the private yet fundamentally social, cooperative and duty-laden character of child-bearing, child-rearing and their bond to the covenant of marriage. Seeking to escape entirely from nature (in order to satisfy a natural desire or a natural right to reproduce!) is self-contradictory in theory and self-alienating in practice. For we are erotic beings only because we are embodied beings, and not merely intellects and wills unfortunately imprisoned in our bodies. And, though health and fitness are clearly great goods, there is something deeply disquieting in looking on our prospective children as artful products perfectible by genetic engineering, increasingly held to our willfully imposed designs, specifications and margins of tolerable error.

    The technical, liberal and meliorist approaches all ignore the deeper anthropological, social and, indeed, ontological meanings of bringing forth new life. To this more fitting and profound point of view, cloning shows itself to be a major alteration, indeed, a major violation, of our given nature as embodied, gendered and engendering beings—and of the social relations built on this natural ground. Once this perspective is recognized, the ethical judgment on cloning can no longer be reduced to a matter of motives and intentions, rights and freedoms, benefits and harms, or even means and ends. It must be regarded primarily as a matter of meaning: Is cloning a fulfillment of human begetting and belonging? Or is cloning rather, as I contend, their pollution and perversion? To pollution and perversion, the fitting response can only be horror and revulsion; and conversely, generalized horror and revulsion are prima facie evidence of foulness and violation. The burden of moral argument must fall entirely on those who want to declare the widespread repugnances of humankind to be mere timidity or superstition.

    Yet repugnance need not stand naked before the bar of reason. The wisdom of our horror at human cloning can be partially articulated, even if this is finally one of those instances about which the heart has its reasons that reason cannot entirely know.

THE PROFUNDITY OF SEX

To see cloning in its proper context, we must begin not, as I did before, with laboratory technique, but with the anthropology—natural and social—of sexual reproduction.

Sexual reproduction—by which I mean the generation of new life from (exactly) two complementary elements, one female, one male, (usually) through coitus—is established (if that is the right term) not by human decision, culture or tradition, but by nature; it is the natural way of all mammalian reproduction. By nature, each child has two complementary biological progenitors. Each child thus stems from and unites exactly two lineages. In natural generation, moreover, the precise genetic constitution of the resulting offspring is determined by a combination of nature and chance, not by human design: each human child shares the common natural human species genotype, each child is genetically (equally) kin to each (both) parent(s), yet each child is also genetically unique.

These biological truths about our origins foretell deep truths about our identity and about our human condition altogether. Every one of us is at once equally human, equally enmeshed in a particular familial nexus of origin, and equally individuated in our trajectory from birth to death—and, if all goes well, equally capable (despite our morality) of participating, with a complementary other, in the very same renewal of such human possibility through procreation. Though less momentous than our common humanity, our genetic individuality is not humanly trivial. It shows itself forth in our distinctive appearance through which we are everywhere recognized; it is revealed in our "signature" marks of fingerprints and our self-recognizing immune system; it symbolizes and foreshadows exactly the unique, never-to-be-repeated character of each human life.

Human societies virtually everywhere have structured child-rearing responsibilities and systems of identity and relationship on the bases of these deep natural facts of begetting. The mysterious yet ubiquitous "love of one's own" is everywhere culturally exploited, to make sure that children are not just produced but well cared for and to create for everyone clear ties of meaning, belonging and obligation. But it is wrong to treat such naturally rooted social practices as mere cultural constructs (like left- or right-driving, or like burying or cremating the dead) that we can alter with little human cost. What would kinship be without its clear natural grounding? And what would identity be

without kinship? We must resist those who have begun to refer to sexual reproduction as the "traditional method of reproduction," who would have us regard
as merely traditional, and by implication arbitrary, what is in truth not only
natural but most certainly profound.

Asexual reproduction, which produces "single-parent" offspring, is a radical
departure from the natural human way, confounding all normal understandings
of father, mother, sibling, grandparent, etc., and all moral relations tied thereto.
It becomes even more of a radical departure when the resulting offspring is a
clone derived not from an embryo, but from a mature adult to whom the clone
would be an identical twin; and when the process occurs not by natural accident (as in natural twinning), but by deliberate human design and manipulation; and when the child's (or children's) genetic constitution is preselected by
the parent(s) (or scientists). Accordingly, as we will see, cloning is vulnerable
to three kinds of concerns and objections, related to these three points: cloning
threatens confusion of identity and individuality, even in small-scale cloning;
cloning represents a giant step (though not the first one) toward transforming
procreation into manufacture, that is, toward the increasing depersonalization
of the process of generation and, increasingly, toward the "production" of
human children as artifacts, products of human will and design (what others
have called the problem of "commodification" of new life); and cloning—like
other forms of eugenic engineering of the next generation—represents a form
of despotism of the cloners over the cloned, and thus (even in benevolent
cases) represents a blatant violation of the inner meaning of parent-child relations, of what it means to have a child, of what it means to say "yes" to our own
demise and "replacement."

Before turning to these specific ethical objections, let me test my claim of
the profundity of the natural way by taking up a challenge recently posed by a
friend. What if the given natural human way of reproduction were asexual, and
we now had to deal with a new technological innovation—artificially induced
sexual dimorphism and the fusing of complementary gametes—whose inventors argued that sexual reproduction promised all sorts of advantages, including
hybrid vigor and the creation of greatly increased individuality? Would one
then be forced to defend natural asexuality because it was natural? Could one
claim that it carried deep human meaning?

The response to this challenge broaches the ontological meaning of sexual
reproduction. For it is impossible, I submit, for there to have been human
life—or even higher forms of animal life—in the absence of sexuality and sexual reproduction. We find asexual reproduction only in the lowest forms of life:
bacteria, algae, fungi, some lower invertebrates. Sexuality brings with it a new
and enriched relationship to the world. Only sexual animals can seek and find
complementary others with whom to pursue a goal that transcends their own

existence. For a sexual being, the world is no longer an indifferent and largely homogeneous *otherness*, in part edible, in part dangerous. It also contains some very special and related and complementary beings, of the same kind but of opposite sex, toward whom one reaches out with special interest and intensity. In higher birds and mammals, the outward gaze keeps a lookout not only for food and predators, but also for prospective mates; the beholding of the many splendored world is suffused with desire for union, the animal antecedent of human eros and the germ of sociality. Not by accident is the human animal both the sexiest animal—whose females do not go into heat but are receptive throughout the estrous cycle and whose males must therefore have greater sexual appetite and energy in order to reproduce successfully—and also the most aspiring, the most social, the most open and the most intelligent animal.

The soul-elevating power of sexuality is, at bottom, rooted in its strange connection to mortality, which it simultaneously accepts and tries to overcome. Asexual reproduction may be seen as a continuation of the activity of self-preservation. When one organism buds or divides to become two, the original being is (doubly) preserved, and nothing dies. Sexuality, by contrast, means perishability and serves replacement; the two that come together to generate one soon will die. Sexual desire, in human beings as in animals, thus serves an end that is partly hidden from, and finally at odds with, the self-serving individual. Whether we know it or not, when we are sexually active we are voting with our genitalia for our own demise. The salmon swimming upstream to spawn and die tell the universal story: sex is bound up with death, to which it holds a partial answer in procreation.

The salmon and the other animals evince this truth blindly. Only the human being can understand what it means. As we learn so powerfully from the story of the Garden of Eden, our humanization is coincident with sexual self-consciousness, with the recognition of our sexual nakedness and all that it implies: shame at our needy incompleteness, unruly self-division and finitude; awe before the eternal; hope in the self-transcending possibilities of children and a relationship to the divine. In the sexually self-conscious animal, sexual desire can become eros, lust can become love. Sexual desire humanly regarded is thus sublimated into erotic longing for wholeness, completion and immortality which drives us knowingly into the embrace and its generative fruit—as well as into all the higher human possibilities of deed, speech and song.

Through children, a good common to both husband and wife, male and female achieve some genuine unification (beyond the mere sexual "union," which fails to do so). The two become one through sharing generous (not needy) love for this third being as good. Flesh of their flesh, the child is the parents' own commingled being externalized, and given a separate and persist-

ing existence. Unification is enhanced also by their commingled work of rearing. Providing an opening to the future beyond the grave, carrying not only our
seed but also our names, our ways and our hopes that they will surpass us in
goodness and happiness, children are a testament to the possibility of transcen-
dence. Gender duality and sexual desire, which first draws our love upward
and outside of ourselves, finally provide for the partial overcoming of the con-
finement and limitation of perishable embodiment altogether.

Human procreation, in sum, is not simply an activity of our rational wills. It
is a more complete activity precisely because it engages us bodily, erotically
and spiritually, as well as rationally. There is wisdom in the mystery of nature
that has joined the pleasure of sex, the inarticulate longing for union, the com-
munication of the loving embrace and the deep-seated and only partly articulate desire for children in the very activity by which we continue the chain of
human existence and participate in the renewal of human possibility. Whether
or not we know it, the severing of procreation from sex, love and intimacy is
inherently dehumanizing, no matter how good the product.

We are now ready for the more specific objections to cloning.

THE PERVERSITIES OF CLONING

First, an important if formal objection: any attempt to clone a human being
would constitute an unethical experiment upon the resulting child-to-be. As
the animal experiments (frog and sheep) indicate, there are grave risk of mishaps and deformities. Moreover, because of what cloning means, one cannot
presume a future cloned child's consent to be a clone, even a healthy one.
Thus, ethically speaking, we cannot even get to know whether or not human
cloning is feasible.

I understand, of course, the philosophical difficulty of trying to compare
a life with defects against nonexistence. Several bioethicists, proud of their
philosophical cleverness, use this conundrum to embarrass claims that one can
injure a child in its conception, precisely because it is only thanks to that com-
plained-of conception that the child is alive to complain. But common sense
tells us that we have no reason to fear such philosophisms. For we surely know
that people can harm and even maim children in the very act of conceiving
them, say, by paternal transmission of the AIDS virus, maternal transmission of
heroin dependence or, arguably, even by bringing them into being as bastards
or with no capacity or willingness to look after them properly. And we believe
that to do this intentionally, or even negligently, is inexcusable and clearly unethical.

The objection about the impossibility of presuming consent may even go
beyond the obvious and sufficient point that a clonant, were he subsequently

to be asked, could rightly resent having been made a clone. At issue are not just benefits and harms, but doubts about the very independence needed to give proper (even retroactive) consent, that is, not just the capacity to choose but the disposition and ability to choose freely and well. It is not at all clear to what extent a clone will truly be a moral agent. For, as we shall see, in the very fact of cloning, and of rearing him as a clone, his makers subvert the cloned child's independence, beginning with that aspect that comes from knowing that one was an unbidden surprise, a gift, to the world, rather than the designed result of someone's artful project.

Cloning creates serious issues of identity and individuality. The cloned person may experience concerns about his distinctive identity not only because he will be in genotype and appearance identical to another human being, but, in this case, because he may also be twin to the person who is his "father" or "mother"—if one can still call them that. What would be the psychic burdens of being the "child" or "parent" of your twin? The cloned individual, moreover, will be saddled with a genotype that has already lived. He will not be fully a surprise to the world. People are likely always to compare his performances in life with that of his alter ego. True, his nurture and his circumstance in life will be different; genotype is not exactly destiny. Still, one must also expect parental and other efforts to shape this new life after the original—or at least to view the child with the original version always firmly in mind. Why else did they clone from the star basketball player, mathematician and beauty queen—or even dear old dad—in the first place?

Since the birth of Dolly, there has been a fair amount of doublespeak on this matter of genetic identity. Experts have rushed in to reassure the public that the clone would in no way be the same person, or have any confusions about his or her identity: as previously noted, they are pleased to point out that the clone of Mel Gibson would not be Mel Gibson. Fair enough. But one is shortchanging the truth by emphasizing the additional importance of the intrauterine environment, rearing and social setting: genotype obviously matters plenty. That, after all, is the only reason to clone, whether human beings or sheep. The odds that clones of Wilt Chamberlain will play in the NBA are, I submit, infinitely greater than they are for clones of Robert Reich.

Curiously, this conclusion is supported, inadvertently, by the one ethical sticking point insisted on by friends of cloning: no cloning without the donor's consent. Though an orthodox liberal objection, it is in fact quite puzzling when it comes from people (such as Ruth Macklin) who also insist that genotype is not identity or individuality, and who deny that a child could reasonably complain about being made a genetic copy. If the clone of Mel Gibson would not be Mel Gibson, why should Mel Gibson have grounds to object that someone had been made his clone? We already allow researchers to use blood and tissue samples for research purposes of no benefit to their sources: my falling hair, my expectorations, my urine and even my biopsied tissues are "not me" and not mine. Courts have held that the profit gained from uses to which scientists put my discarded tissues do not legally belong to me. Why, then, no cloning without consent—including, I assume, no cloning from the body of someone who just died? What harm is done the donor, if genotype is "not me"? Truth to tell, the only powerful justification for objecting is that genotype really does have something to do with identity, and everybody knows it. If not, on what basis could Michael Jordan object that someone cloned "him," say, from cells taken from a "lost" scraped-off piece of his skin? The insistence on donor consent unwittingly reveals the problem of identity in all cloning.

Genetic distinctiveness not only symbolizes the uniqueness of each human life and the independence of its parents that each human child rightfully attains. It can also be an important support for living a worthy and dignified life. Such arguments apply with great force to any large-scale replication of human individuals. But they are sufficient, in my view, to rebut even the first attempts to clone a human being. One must never forget that these are human beings upon whom our eugenic or merely playful fantasies are to be enacted.

Troubled psychic identity (distinctiveness), based on all-too-evident genetic identity (sameness), will be made much worse by the utter confusion of social identity and kinship ties. For, as already noted, cloning radically confounds lineage and social relations, for "offspring" as for "parents." As bioethicist James Nelson has pointed out, a female child cloned from her "mother" might develop a desire for a relationship to her "father," and might understandably seek out the father of her "mother," who is after all also her biological twin sister. Would "grandpa," who thought his paternal duties concluded, be pleased to discover that the clonant looked to him for paternal attention and support?

Social identity and social ties of relationship and responsibility are widely connected to, and supported by, biological kinship. Social taboos on incest (and adultery) everywhere serve to keep clear who is related to whom (and especially which child belongs to which parents), as well as to avoid confounding the social identity of parent-and-child (or brother-and-sister) with the social identity of lovers, spouses and co-parents. True, social identity is altered by adoption (but as a matter of the best interest of already living children: we do not deliberately produce children for adoption). True, artificial insemination and in vitro fertilization with donor sperm, or whole embryo donation, are in some way forms of "prenatal adoption"—a not altogether unproblematic practice. Even here, though, there is in each case (as in all sexual reproduction) a known male source of sperm and a known single female source of egg—a genetic father and a genetic mother—should anyone care to know (as adopted children often do) who is genetically related to whom.

In the case of cloning, however, there is but one "parent." The usually sad situation of the "single-parent child" is here deliberately planned, and with a vengeance. In the case of self-cloning, the "offspring" is, in addition, one's twin; and so the dreaded result of incest—to be parent to one's sibling—is here brought about deliberately, albeit without any act of coitus. Moreover, all other relationships will be confounded. What will father, grandfather, aunt, cousin, sister mean? Who will bear what ties and what burdens? What sort of social identity will someone have with one whole side—"father's" or "mother's"—necessarily excluded? It is no answer to say that our society, with its high incidence of divorce, remarriage, adoption, extramarital childbearing and the rest, already confounds lineage and confuses kinship and responsibility for children (and everyone else), unless one also wants to argue that this is, for children, a preferable state of affairs.

Human cloning would also represent a giant step toward turning begetting into making, procreation into manufacture (literally, something "handmade"), a process already begun with in vitro fertilization and genetic testing of embryos. With cloning, not only is the process in hand, but the total genetic blueprint of the cloned individual is selected and determined by the human artisans. To be sure, subsequent development will take place according to natural processes; and the resulting children will still be recognizably human. But we here would be taking a major step into making man himself simply another one of the man-made things. Human nature becomes merely the last part of nature to succumb to the technological project, which turns all of nature into raw material at human disposal, to be homogenized by our rationalized technique according to the subjective prejudices of the day.

How does begetting differ from making? In natural procreation, human beings come together, complementarily male and female, to give existence to another being who is formed, exactly as we were, *by what we are*: living, hence perishable, hence aspiringly erotic, human beings. In clonal reproduction, by contrast, and in the more advanced forms of manufacture to which it leads, we give existence to a being not by what we are but by what we intend and design. As with any product of our making, no matter how excellent, the artificer stands above it, not as an equal but as a superior, transcending it by his will and creative prowess. Scientists who clone animals make it perfectly clear that they are engaged in instrumental making; the animals are, from the start, designed as means to serve rational human purposes. In human cloning, scientists and prospective "parents" would be adopting the same technocratic mentality to human children: human children would be their artifacts.

Such an arrangement is profoundly dehumanizing, no matter how good the product. Mass-scale cloning of the same individual makes the point vividly; but the violation of human equality, freedom and dignity are present even in a single planned clone. And procreation dehumanized into manufacture is further degraded by commodification, a virtually inescapable result of allowing baby-making to proceed under the banner of commerce. Genetic and reproductive biotechnology companies are already growth industries, but they will go into commercial orbit once the Human Genome Project nears completion. Supply will create enormous demand. Even before the capacity for human cloning arrives, established companies will have invested in the harvesting of eggs from ovaries obtained at autopsy or through ovarian surgery, practiced embryonic genetic alteration, and initiated the stockpiling of prospective donor tissues. Through the rental of surrogate-womb services, and through the buying and selling of tissues and embryos, priced according to the merit of the donor, the commodification of nascent human life will be unstoppable.

Finally, and perhaps most important, the practice of human cloning by nuclear transfer—like other anticipated forms of genetic engineering of the next generation—would enshrine and aggravate a profound and mischievous misunderstanding of the meaning of having children and of the parent-child relationship. When a couple now chooses to procreate, the partners are saying yes to the emergence of new life in its novelty, saying yes not only to having a child but also, tacitly, to having whatever child this child turns out to be. In accepting our finitude and opening ourselves to our replacement, we are tacitly confessing the limits of our control. In this ubiquitous way of nature, embracing the future by procreating means precisely that we are relinquishing our grip, in the very activity of taking up our own share in what we hope will be the immortality of human life and the human species. This means that our children are not *our* children: they are not our property, not our possessions. Neither are they supposed to live our lives for us, or anyone else's life but their own. To be sure, we seek to guide them on their way, imparting to them not just life but nurturing, love, and a way of life; to be sure, they bear our hopes that they will live fine and flourishing lives, enabling us in small measure to transcend our own limitations. Still, their genetic distinctiveness and independence are the natural foreshadowing of the deep truth that they have their own and never-before-enacted life to live. They are sprung from a past, but they take an uncharted course into the future.

Much harm is already done by parents who try to live vicariously through their children. Children are sometimes compelled to fulfill the broken dreams of unhappy parents; John Doe Jr. or the III is under the burden of having to live up to his forebear's name. Still, if most parents have hopes for their children, cloning parents will have expectations. In cloning, such overbearing parents take at the start a decisive step which contradicts the entire meaning of the open and forward-looking nature of parent-child relations. The child is given a genotype that has already lived, with full expectation that this blueprint of a past life ought to be controlling of the life that is to come. Cloning is inherently despotic, for it seeks to make one's children (or someone else's children) after one's own image (or an image of one's choosing) and their future according to one's will. In some cases, the despotism may be mild and benevolent. In other cases, it will be mischievous and downright tyrannical. But despotism—the control of another through one's will—it inevitably will be.

MEETING SOME OBJECTIONS

The defenders of cloning, of course, are not wittingly friends of despotism. Indeed, they regard themselves mainly as friends of freedom: the freedom of individuals to reproduce, the freedom of scientists and inventors to discover and devise and to foster "progress" in genetic knowledge and technique. They want large-scale cloning only for animals, but they wish to preserve cloning as a human option for exercising our "right to reproduce"—our right to have children, and children with "desirable genes." As law professor John Robertson points out, under our "right to reproduce" we already practice early forms of unnatural, artificial and extramarital reproduction, and we already practice early forms of eugenic choice. For this reason, he argues, cloning is no big deal.

We have here a perfect example of the logic of the slippery slope, and the slippery way in which it already works in this area. Only a few years ago, slippery slope arguments were used to oppose artificial insemination and in vitro fertilization using unrelated sperm donors. Principles used to justify these practices, it was said, will be used to justify more artificial and more eugenic practices, including cloning. Not so, the defenders retorted, since we can make the necessary distinctions. And now, without even a gesture at making the necessary distinctions, the continuity of practice is held by itself to be justificatory.

The principle of reproductive freedom as currently enunciated by the proponents of cloning logically embraces the ethical acceptability of sliding down the entire rest of the slope—to producing children ectogenetically from sperm to term (should it become feasible) and to producing children whose entire genetic makeup will be the product of parental eugenic planning and choice. If reproductive freedom means the right to have a child of one's own choosing, by whatever means, it knows and accepts no limits.

But, far from being legitimated by a "right to reproduce," the emergence of techniques of assisted reproduction and genetic engineering should compel us to reconsider the meaning and limits of such a putative right. In truth, a "right to reproduce" has always been a peculiar and problematic notion. Rights generally belong to individuals, but this is a right which (before cloning) no one can exercise alone. Does the right then inhere only in couples? Only in married couples? Is it a (woman's) right to carry or deliver or a right (of one or more parents) to nurture and rear? Is it a right to have your own biological child? Is it a right only to attempt reproduction, or a right also to succeed? Is it a right to acquire the baby of one's choice?

The assertion of a negative "right to reproduce" certainly makes sense when it claims protection against state interference with procreative liberty, say, through a program of compulsory sterilization. But surely it cannot be the basis of a tort claim against nature, to be made good by technology, should free efforts at natural procreation fail. Some insist that the right to reproduce embraces also the right against state interference with the free use of all technological means to obtain a child. Yet such a position cannot be sustained: for reasons having to do with the means employed, any community may rightfully prohibit surrogate pregnancy, or polygamy, or the sale of babies to infertile couples, without violating anyone's basic human "right to reproduce." When the exercise of a previously innocuous freedom now involves or impinges on troublesome practices that the original freedom never was intended to reach, the general presumption of liberty needs to be reconsidered.

We do indeed already practice negative eugenic selection, through genetic screening and prenatal diagnosis. Yet our practices are governed by a norm of health. We seek to prevent the birth of children who suffer from known (serious) genetic diseases. When and if gene therapy becomes possible, such diseases could then be treated, in utero or even before implantation—I have no ethical objection in principle to such a practice (though I have some practical worries), precisely because it serves the medical goal of healing existing individuals. But therapy, to be therapy, implies not only an existing "patient." It also implies a norm of health. In this respect, even germline gene "therapy," though practiced not on a human being but on egg and sperm, is less radical than cloning, which is in no way therapeutic. But once one blurs the distinction between health promotion and genetic enhancement, between so-called negative and positive eugenics, one opens the door to all future eugenic designs. "To make sure that a child will be healthy and have good chances in life": this is Robertson's principle, and owing to its latter clause it is an utterly elastic principle, with no boundaries. Being over eight feet tall will likely produce some very good chances in life, and so will having the looks of Marilyn Monroe, and so will a genius-level intelligence.

Proponents want us to believe that there are legitimate uses of cloning that can be distinguished from illegitimate uses, but by their own principles no such limits can be found. (Nor could any such limits be enforced in practice.) Reproductive freedom, as they understand it, is governed solely by the subjective wishes of the parents-to-be (plus the avoidance of bodily harm to the child). The sentimentally appealing case of the childless married couple is, on these grounds, indistinguishable from the case of an individual (married or not) who would like to clone someone famous or talented, living or dead. Further, the principle here endorsed justifies not only cloning but, indeed, all future artificial attempts to create (manufacture) "perfect" babies.

A concrete example will show how, in practice no less than in principle, the so-called innocent case will merge with, or even turn into, the more troubling ones. In practice, the eager parents-to-be will necessarily be subject to the tyranny of expertise. Consider an infertile married couple, she lacking eggs or he lacking sperm, that wants a child of their (genetic) own, and propose to clone either husband or wife. The scientist-physician (who is also coowner of the cloning company) points out the likely difficulties—a cloned child is not really their (genetic) child, but the child of only *one* of them; this imbalance may produce strains on the marriage; the child might suffer identity confusion; there is a risk of perpetuating the cause of sterility; and so on—and he also points out the advantages of choosing a donor nucleus. Far better than a child of their own would be a child of their own choosing. Touting his own expertise in selecting healthy and talented donors, the doctor presents the couple with his latest catalog containing the pictures, the health records and the accomplishments of his stable of cloning donors, samples of whose tissues are in his deep freeze. Why not, dearly beloved, a more perfect baby?

The "perfect baby," of course, is the project not of the infertility doctors, but of the eugenic scientists and their supporters. For them, the paramount right is not the so-called right to reproduce but what biologist Bentley Glass called, a quarter of a century ago, "the right of every child to be born with a sound physical and mental constitution, based on a sound genotype . . . the inalienable right to a sound heritage." But to secure this right, and to achieve the requisite quality control over new human life, human conception and gestation will need to be brought fully into the bright light of the laboratory, beneath which it can be fertilized, nourished, pruned, weeded, watched, inspected, prodded, pinched, cajoled, injected, tested, rated, graded, approved, stamped, wrapped, sealed and delivered. There is no other way to produce the perfect baby.

Yet we are urged by proponents of cloning to forget about the science fiction scenarios of laboratory manufacture and multiple-copied clones, and to focus only on the homely cases of infertile couples exercising their reproductive rights. But why, if the single cases are so innocent, should multiplying their performance be so off-putting? (Similarly, why do others object to people making money off this practice, if the practice itself is perfectly acceptable?) When we follow the sound ethical principle of universalizing our choice—"would it be right if everyone cloned a Wilt Chamberlain (with his consent, of course)? Would it be right if everyone decided to practice asexual reproduction?"—we discover what is wrong with these seemingly innocent cases. The so-called science fiction cases make vivid the meaning of what looks to us, mistakenly, to be benign.

Though I recognize certain continuities between cloning and, say, in vitro fertilization, I believe that cloning differs in essential and important ways. Yet those who disagree should be reminded that the "continuity" argument cuts both ways. Sometimes we establish bad precedents, and discover that they were bad only when we follow their inexorable logic to places we never meant to go. Can the defenders of cloning show us today how, on their principles, we will be able to see producing babies ("perfect babies") entirely in the laboratory or exercising full control over their genotypes (including so-called enhancement) as ethically different, in any essential way, from present forms of assisted reproduction? Or are they willing to admit, despite their attachment to the principle of continuity, that the complete obliteration of "mother" or "father," the complete depersonalization of procreation, the complete manufacture of human beings and the complete genetic control of one generation over the next would be ethically problematic and essentially different from current forms of assisted reproduction? If so, where and how will they draw the line, and why? I draw it at cloning, for all the reasons given.

BAN THE CLONING OF HUMANS

What, then, should we do? We should declare that human cloning is unethical in itself and dangerous in its likely consequences. In so doing, we shall have the backing of the overwhelming majority of our fellow Americans, and of the human race, and (I believe) of most practicing scientists. Next, we should do all that we can to prevent the cloning of human beings. We should do this by means of an international legal ban if possible, and by a unilateral national ban, at a minimum. Scientists may secretly undertake to violate such a law, but they will be deterred by not being able to stand up proudly to claim the credit for their technological bravado and success. Such a ban on clonal baby-making, moreover, will not harm the progress of basic genetic science and technology. On the contrary, it will reassure the public that scientists are happy to proceed without violating the deep ethical norms and intuitions of the human community.

This still leaves the vexed question about laboratory research using early embryonic human clones, specially created only for such research purposes, with no intention to implant them into a uterus. There is no question that such research holds great promise for gaining fundamental knowledge about normal (and abnormal) differentiation, and for developing tissue lines for transplantation that might be used, say, in treating leukemia or in repairing brain or spinal cord injuries—to mention just a few of the conceivable benefits. Still, unrestricted clonal embryo research will surely make the production of living human clones much more likely. Once the genies put the cloned embryos into the bottles, who can strictly control where they go (especially in the absence of legal prohibitions against implanting them to produce a child)?

I appreciate the potentially great gains in scientific knowledge and medical treatment available from embryo research, especially with cloned embryos. At the same time, I have serious reservations about creating human embryos for the sole purpose of experimentation. There is something deeply repugnant and fundamentally transgressive about such a utilitarian treatment of prospective human life. This total, shameless exploitation is worse, in my opinion, than the "mere" destruction of nascent life. But I see no added objections, as a matter of principle, to creating and using *cloned* early embryos for research purposes, beyond the objections that I might raise to doing so with embryos produced sexually.

And yet, as a matter of policy and prudence, any opponent of the manufacture of cloned humans must, I think, in the end oppose also the creating of cloned human embryos. Frozen embryonic clones (belonging to whom?) can be shuttled around without detection. Commercial ventures in human cloning will be developed without adequate oversight. In order to build a fence around the law, prudence dictates that one oppose—for this reason alone—all production of cloned human embryos, even for research purposes. We should allow for all cloning research on animals to go forward, but the only safe trench that we can dig across the slippery slope, I suspect, is to insist on the inviolable distinction between animal and human cloning.

Some readers, and certainly most scientists, will not accept such prudent restraints, since they desire the benefits of research. They will prefer, even in fear and trembling, to allow human embryo cloning research to go forward.

Very well. Let us test them. If the scientists want to be taken seriously on ethical grounds, they must at the very least agree that embryonic research may proceed if and only if it is preceded by an absolute and effective ban on all attempts to implant into a uterus a cloned human embryo (cloned from an adult) to produce a living child. Absolutely no permission for the former without the latter.

The National Bioethics Advisory Commission's recommendations regarding this matter should be watched with the greatest care. Yielding to the wishes of the scientists, the commission will almost surely recommend that cloning human embryos for research be permitted. To allay public concern, it will likely also call for a temporary moratorium—not a legislative ban—on implanting cloned embryos to make a child, at least until such time as cloning techniques will have been perfected and rendered "safe" (precisely through the permitted research with cloned embryos). But the call for a moratorium rather than a legal ban would be a moral and a practical failure. Morally, this ethics commission would (at best) be waffling on the main ethical question, by refusing to declare the production of human clones unethical (or ethical). Practically, a moratorium on implantation cannot provide even the minimum protection needed to prevent the production of cloned humans.

Opponents of cloning need therefore to be vigilant. Indeed, no one should be willing even to consider a recommendation to allow the embryo research to proceed unless it is accompanied by a call for *prohibiting* implantation and until steps are taken to make such a prohibition effective.

Technically, the National Bioethics Advisory Commission can advise the president only on federal policy, especially federal funding policy. But given the seriousness of the matter at hand, and the grave public concern that goes beyond federal funding, the commission should take a broader view. (If it doesn't, Congress surely will.) Given that most assisted reproduction occurs in the private sector, it would be cowardly and insufficient for the commission to say, simply, "no federal funding" for such practices. It would be disingenuous to argue that we should allow federal funding so that we would then be able to regulate the practice; the private sector will not be bound by such regulations. Far better, for virtually everyone concerned, would be to distinguish between research on embryos and baby-making, and to call for a complete national and international ban (effected by legislation and treaty) of the latter, while allowing the former to proceed (at least in private laboratories).

The proposal for such a legislative ban is without American precedent, at least in technological matters, though the British and others have banned cloning of human beings, and we ourselves ban incest, polygamy and other forms of "reproductive freedom." Needless to say, working out the details of such a ban, especially a global one, would be tricky, what with the need to develop appropriate sanctions for violators. Perhaps such a ban will prove ineffective; perhaps it will eventually be shown to have been a mistake. But it would at least place the burden of practical proof where it belongs: on the proponents of this horror, requiring them to show very clearly what great social or medical good can be had only by the cloning of human beings.

We Americans have lived by, and prospered under, a rosy optimism about scientific and technological progress. The technological imperative—if it can be done, it must be done—has probably served us well, though we should admit that there is no accurate method for weighing benefits and harms. Even when, as in the cases of environmental pollution, urban decay or the lingering deaths that are the unintended by-products of medical success, we recognize the unwelcome outcomes of technological advance, we remain confident in our ability to fix all the "bad" consequences—usually by means of still newer and better technologies. How successful we can continue to be in such post hoc repairing is at least an open question. But there is very good reason for shifting the paradigm around, at least regarding those technological interventions into the human body and mind that will surely effect fundamental (and likely irreversible) changes in human nature, basic human relationships, and what it means to be a human being. Here we surely should not be willing to risk everything in the naïve hope that, should things go wrong, we can later set them right.

The president's call for a moratorium on human cloning has given us an important opportunity. In a truly unprecedented way, we can strike a blow for the human control of the technological project, for wisdom, prudence and human dignity. The prospect of human cloning, so repulsive to contemplate, is the occasion for deciding whether we shall be slaves of unregulated progress, and ultimately its artifacts, or whether we shall remain free human beings who guide our technique toward the enhancement of human dignity. If we are to seize the occasion, we must, as the late Paul Ramsey wrote,

> raise the ethical questions with a serious and not a frivolous conscience. A man of frivolous conscience announces that there are ethical quandaries ahead that we must urgently consider before the future catches up with us. By this he often means that we need to devise a new ethics that will provide the rationalization for doing in the future what men are bound to do because of new actions and interventions science will have made possible. In contrast a man of serious conscience means to say in raising urgent ethical questions that there may be some things that men should never do. The good things that men do can be made complete only by the things they refuse to do.

BEGETTING AND

CLONING

GILBERT MEILAENDER

Gilbert Meilaender has written frequently about bioethics over the last twenty years, especially opposing the withdrawal of feeding tubes from comatose patients. He writes as a Lutheran theologian.

In this piece, "Begetting and Cloning," Meilaender goes back to Genesis in the Old Testament to argue that there is something divinely inspired about the way humans are made through sex and that to allow another form of human reproduction cheapens God's grace and the idea of a child as a gift. Meilaender opposes the creation of children motivated by human will alone. He opposes allowing human cloning in order to retain the mystery of human creation.

Gilbert Meilaender holds the Board of Directors Chair in theological ethics at Valparaiso University. He was formerly dean of Oberlin College.

(The following remarks were presented to the National Bioethics Advisory Commission on March 13, 1997.)

I HAVE BEEN INVITED, AS I UNDERSTAND IT, TO SPEAK TODAY SPE-
cifically as a Protestant theologian. I have tried to take that charge seriously, and I have chosen my concerns accordingly. I do not suppose, therefore, that the issues I address are the only issues to which you ought to give your attention. Thus, for example, I will not address the question of whether we could rightly conduct the first experiments in human cloning, given the likelihood that such experiments would not at first fully succeed. That is an important moral question, but I will not take it up. Nor do I suppose that I can represent Protestants generally. There is no such beast. Indeed, Protestants are specialists in the art of fragmentation. In my own tradition, which is Lutheran, we

commonly understand ourselves as quite content to be Catholic except when, on certain questions, we are compelled to disagree. Other Protestants might think of themselves differently.

More important, however, is this point: Attempting to take my charge seriously, I will speak theologically—not just in the standard language of bioethics or public policy. I do not think of this, however, simply as an opportunity for the "Protestant interest group" to weigh in at your deliberations. On the contrary, this theological language has sought to uncover what is universal and human. It begins epistemologically from a particular place, but it opens up ontologically a vision of the human. The unease about human cloning that I will express is widely shared. I aim to get at some of the theological underpinnings of that unease in language that may seem unfamiliar or even unwelcome, but it is language that is grounded in important Christian affirmations that seek to understand the child as our equal—one who is a gift and not a product. In any case, I will do you the honor of assuming that you are interested in hearing what those who speak such a language have to say, and I will also suppose that a faith which seeks understanding may sometimes find it.

Lacking an accepted teaching office within the church, Protestants had to find some way to provide authoritative moral guidance. They turned from the authority of the church as interpreter of Scripture to the biblical texts themselves. That characteristic Protestant move is not likely, of course, to provide any very immediate guidance on a subject such as human cloning. But it does teach something about the connection of marriage and parenthood. The creation story in the first chapter of Genesis depicts the creation of humankind as male and female, sexually differentiated and enjoined by God's grace to sustain human life through procreation.

Hence, there is given in creation a connection between the differentiation of the sexes and the begetting of a child. We begin with that connection, making our way indirectly toward the subject of cloning. It is from the vantage point of this connection that our theological tradition has addressed two questions that are both profound and mysterious in their simplicity: What is the meaning of a child? And what is good for a child? These questions are, as you know, at the heart of many problems in our society today, and it is against the background of such questions that I want to reflect upon the significance of human cloning. What Protestants found in the Bible was a normative view: namely, that the sexual differentiation is ordered toward the creation of offspring, and children should be conceived within the marital union. By God's grace the child is a gift who springs from the giving and receiving of love. Marriage and parenthood are connected—held together in a basic form of humanity.

To this depiction of the connection between sexual differentiation and childbearing as normative, it is, as Anglican theologian Oliver O'Donovan has argued, possible to respond in different ways. We may welcome the connection and find in it humane wisdom to guide our conduct. We may resent it as a limit to our freedom and seek to transcend it. We did not need modern scientific breakthroughs to know that it is possible—and sometimes seemingly desirable—to sever the connection between marriage and begetting children. The possibility of human cloning is striking only because it breaks the connection so emphatically. It aims directly at the heart of the mystery that is a child. Part of the mystery here is that we will always be hard-pressed to explain why the connection of sexual differentiation and procreation should not be broken. Precisely to the degree that it is a basic form of humanity, it will be hard to give more fundamental reasons why the connection should be welcomed and honored when, in our freedom, we need not do so. But moral argument must begin somewhere. To see through everything is, as C. S. Lewis once put it, the same as not to see at all.

If we cannot argue to this starting point, however, we can argue from it. If we cannot entirely explain the mystery, we can explicate it. And the explication comes from two angles. Maintaining the connection between procreation and the sexual relationship of a man and woman is good both for that relationship and for children.

It is good, first, for the relation of the man and woman. No doubt the motives of those who beget children coitally are often mixed, and they may be uncertain about the full significance of what they do. But if they are willing to shape their intentions in accord with the norm I have outlined, they may be freed from self-absorption. The act of love is not simply a personal project undertaken to satisfy one's own needs, and procreation, as the fruit of coitus, reminds us of that. Even when the relation of a man and woman does not or cannot give rise to offspring, they can understand their embrace as more than their personal project in the world, as their participation in a form of life that carries its own inner meaning and has its telos established in the creation. The meaning of what we do then is not determined simply by our desire or will. As Oliver O'Donovan has noted, some understanding like this is needed if the sexual relation of a man and woman is to be more than "simply a profound form of play."

And when the sexual act becomes only a personal project, so does the child. No longer then is the bearing and rearing of children thought of as a task we should take up or as a return we make for the gift of life; instead, it is a project we undertake if it promises to meet our needs and desires. Those people—both learned commentators and ordinary folk—who in recent days have described cloning as narcissistic or as replication of one's self see something important.

Even if we grant that a clone, reared in different circumstances than its immediate ancestor, might turn out to be quite a different person in some respects,
the point of that person's existence would be grounded in our will and desire.

Hence, retaining the tie that unites procreation with the sexual relation of a
man and woman is also good for children. Even when a man and woman deeply
desire a child, the act of love itself cannot take the child as its primary object.
They must give themselves to each other, setting aside their projects, and the
child becomes the natural fruition of their shared love—something quite different from a chosen project. The child is therefore always a gift—one like
them who springs from their embrace, not a being whom they have made and
whose destiny they should determine. This is light years away from the notion
that we all have a right to have children—in whatever way we see fit, whenever
it serves our purposes. Our children begin with a kind of genetic independence
of us, their parents. They replicate neither their father nor their mother. That
is a reminder of the independence that we must eventually grant to them and
for which it is our duty to prepare them. To lose, even in principle, this sense
of the child as a gift entrusted to us will not be good for children.

I will press this point still further by making one more theological move. When
Christians tried to tell the story of Jesus as they found it in their Scriptures,
they were driven to some rather complex formulations. They wanted to say
that Jesus was truly one with that God whom he called Father, lest it should
seem that what he had accomplished did not really overcome the gulf that
separates us from God. Thus, while distinguishing the persons of Father and
Son, they wanted to say that Jesus is truly God—of one being with the Father.
And the language in which they did this (in the fourth-century Nicene Creed,
one of the two most important creeds that antedate the division of the church
in the West at the Reformation) is language which describes the Son of the
Father as "begotten, not made." Oliver O'Donovan has noted that this distinction between making and begetting, crucial for Christians' understanding of
God, carries considerable moral significance.

What the language of the Nicene Creed wanted to say was that the Son is
God just as the Father is God. It was intended to assert an equality of being.
And for that what was needed was a language other than the language of making. What we beget is like ourselves. What we make is not; it is the product of
our free decision, and its destiny is ours to determine. Of course, on this Christian understanding human beings are not begotten in the absolute sense that
the Son is said to be begotten of the Father. They are made—but made by
God through human begetting. Hence, although we are not God's equal, we
are of equal dignity with each other. And we are not at each other's disposal.

If it is, in fact, human begetting that expresses our equal dignity, we should not lightly set it aside in a manner as decisive as cloning.

I am well aware, of course, that other advances in what we are pleased to call reproductive technology have already strained the connection between the sexual relationship of a man and woman and the birth of a child. Clearly, procreation has to some extent become reproduction, making rather than doing. I am far from thinking that all this has been done well or wisely, and sometimes we may only come to understand the nature of the road we are on when we have already traveled fairly far along it. But whatever we say of that, surely human cloning would be a new and decisive turn on this road—far more emphatically a kind of production, far less a surrender to the mystery of the genetic lottery which is the mystery of the child who replicates neither father nor mother but incarnates their union, far more an understanding of the child as a product of human will.

I am also aware that we can all imagine circumstances in which we ourselves might—were the technology available—be tempted to turn to cloning. Parents who lose a young child in an accident and want to "replace" her. A seriously ill person in need of embryonic stem cells to repair damaged tissue. A person in need of organs for transplant. A person who is infertile and wants, in some sense, to reproduce. Once the child becomes a project or product, such temptations become almost irresistible. There is no end of good causes in the world, and they would sorely tempt us even if we did not live in a society for which the pursuit of health has become a god, justifying almost anything.

As theologian and bioethicist William F. May has often noted, we are preoccupied with death and the destructive powers of our world. But without in any way glorifying suffering or pretending that it is not evil, Christians worship a God who wills to be with us in our dependence, teaching us "attentiveness before a good and nurturant God." We learn therefore that what matters is how we live, not only how long—that we are responsible to do as much good as we can, but this means, as much as we can within the limits morality sets for us.

I am also aware, finally, that we might for now approve human cloning but only in restricted circumstances—as, for example, the cloning of preimplantation embryos (up to fourteen days) for experimental use. That would, of course, mean the creation solely for purposes of research of human embryos—human subjects who are not really best described as preimplantation embryos. They are unimplanted embryos—a locution that makes clear the extent to which their being and destiny are the product of human will alone. If we are genuinely baffled about how best to describe the moral status of that human subject who is the unimplanted embryo, we should not go forward in a way that peculiarly combines metaphysical bewilderment with practical certitude by approving even such limited cloning for experimental purposes.

Protestants are often pictured—erroneously in many respects—as stout defenders of human freedom. But whatever the accuracy of that depiction, they have not had in mind a freedom without limit, without even the limit that is God. They have not located the dignity of human beings in a self-modifying freedom that knows no limit and that need never respect a limit which it can, in principle, transgress. It is the meaning of the child—offspring of a man and woman, but a replication of neither; their offspring, but not their product whose meaning and destiny they might determine—that, I think, constitutes such a limit to our freedom to make and remake ourselves. In the face of that mystery I hope that your Commission will remember that "progress" is always an optional goal in which nothing of the sacred inheres.

CLONING

HUMAN BEINGS

NATIONAL BIOETHICS ADVISORY COMMISSION

Several days after the announcement of the origination of Dolly by cloning, and after a world-wide frenzy of speculation about human cloning, President Clinton asked a little-known, fifteen-member bioethics commission to investigate American public policy on this topic. It was also asked to hold hearings so that various groups could testify.

The National Bioethics Advisory Commission (NBAC) had two serious limitations: First it was given only ninety days to issue its conclusions; second, its members had not been appointed for their expertise on medicine and reproductive ethics. Instead, the original mandate of NBAC was investigation of genetic privacy and the ethics of medical research.

The following selection reprints two small sections from the final NBAC report, Cloning Human Beings. In the first, NBAC outlines the ethical reasons it has heard against cloning human beings (it distinguishes these from religious reasons). NBAC does not necessarily endorse these reasons, it merely presents the reasons that it decided were worth analyzing. The second part reprints the actual conclusions of NBAC about human cloning.

The NBAC conclusions were controversial. At present, experimental forms of human reproduction in fertility clinics are not regulated by the federal government or most state governments. NBAC could have called for regulation of human cloning in such clinics, either by state or federal agencies. Another option would have been to call for a ban on federal funding of human cloning while allowing it to occur privately—as now occurs with research on human embryos.

NBAC went much further and called on Congress to pass a law making it a federal crime to originate a child by somatic cell nuclear transfer. This would be a gigantic change in the climate of assisted reproduction in medicine and

*would represent a radical move by the federal government into what has tradi-
tionally been the personal reproductive lives of families.*

ETHICAL CONSIDERATIONS

*The prospect of creating children through somatic cell nuclear transfer has
elicited widespread concern, much of it in the form of fears about harms to the
children who may be born as a result. There are concerns about possible physical harms from the manipulations of ova, nuclei, and embryos which are parts
of the technology, and also about possible psychological harms, such as a diminished sense of individuality and personal autonomy. There are ethical concerns
as well about a degradation of the quality of parenting and family life if parents
are tempted to seek excessive control over their children's characteristics, to
value children according to how well they meet overly detailed parental expectations, and to undermine the acceptance and openness that typify loving families. Virtually all people agree that the current risks of physical harm to
children associated with somatic cell nuclear transplantation cloning justify a
prohibition at this time on such experimentation. In addition to concerns about
specific harms to children, people have frequently expressed fears that a widespread practice of such cloning would undermine important social values, such
as opening the door to a form of eugenics or by tempting some to manipulate
others as if they were objects instead of persons, and exceeding the moral
boundaries inherent in the human condition. Arrayed against these concerns
are other important social values, such as protecting personal choice, maintaining privacy and the freedom of scientific inquiry, and encouraging the possible
development of new biomedical breakthroughs. As somatic cell nuclear transfer
cloning could represent a means of human reproduction for some people, limitations on that choice must be made only when the societal benefits of prohibition clearly outweigh the value of maintaining the private nature of such highly
personal decisions. Especially in light of some arguably compelling cases for
attempting to create a child through somatic cell nuclear transfer, the ethics of
policy making must strike a balance between the values we, as a society, wish
to reflect and the freedom of individual choice and any liberties we propose to
limit.*

One of the key challenges for the Commission has been to understand many
of the moral and religious objections to creating human beings using somatic
cell nuclear transfer as well as to investigate and articulate the widespread
intuitive disapproval of cloning human beings in this manner.[1] This challenge
included an initial attempt to examine the plausibility and persuasiveness of

these objections and of the counter arguments or specially compelling and specific cases for deploying this technology. As with the concerns offered in opposition to cloning, those offered in its defense also must be examined for their plausibility and persuasiveness. Religious perspectives were presented in the previous chapter. This chapter focuses on ethical principles not tied to any particular religious tradition, although these broad principles may be incorporated in the teachings of many religions.

The task is made quite difficult by the fact that neither moral philosophers nor religious thinkers can agree on the "best" moral theory; indeed, they often cannot even agree on the practical implications of any single theory. For example, some people base their arguments on an assessment of the particular harms and benefits that would flow to individuals and to society if somatic cell nuclear transfer techniques were to become commonplace. Others express their views by arguing about overarching rights—the child's right to individuality and dignity versus the nucleus donor's right to procreate or the scientist's right to do research. And while moral and even human rights are not necessarily understood as absolute, a choice to violate such rights requires more than a simple balancing of benefits over harms.

While some of the risks and benefits of somatic cell nuclear cloning of human beings are well enough understood to support the conclusion that it should not be permitted at this time, the difficult task of striking the balance among competing rights and interests needs more time for discussion and development. This chapter reviews some of these arguments which may serve as the starting point for a profound and sustained reflection on the significance of creating children through somatic cell nuclear transfer.

The following discussion of issues raised by such cloning begins with an important caveat. Any research or clinical experiment on creating a child in this manner would involve the creation of an embryo. That is, the fusion of a human somatic cell and an egg whose nucleus has been removed would produce a human embryo, with the apparent potential to be implanted *in utero* and developed to term. Ethical concerns surrounding the issues of embryo research, absent the implantation and carrying to term of an embryo, have recently received extensive analysis and deliberation in our country (National Institutes of Health, 1994). Indeed, as described in Chapter Five, federal funding for human embryo research is severely restricted, although there are few restrictions on human embryo research carried out in the private sector using non-federal funds.

The unique prospect, vividly raised by Dolly, is the creation of a new individual genetically identical to an existing (or previously existing) person—a "delayed" genetic twin. This prospect has been the source of the overwhelming public concern about such cloning. While the creation of embryos for research purposes alone always raises serious ethical questions, the use of somatic cell nuclear transfer to create embryos raises no new issues in this respect. The unique and distinctive ethical issues raised by the use of somatic cell nuclear transfer to create children relate to, for example, serious safety concerns, individuality, family integrity, and treating children as objects. Consequently, the Commission focused its attention on the use of such techniques for the purpose of creating an embryo which would then be implanted in a woman's uterus and brought to term. It also expanded its analysis of this particular issue to encompass activities in both the public and private sector.

Potential for Physical Harms

There is one basis of opposition to somatic cell nuclear transfer cloning on which almost everyone can agree. For reasons outlined in Chapter Two, there is virtually universal concern regarding the current safety of attempting to use this technique in human beings. Even if there were a compelling case in favor of creating a child in this manner, it would have to yield to one fundamental principle of both medical ethics and political philosophy—the injunction, as it is stated in the Hippocratic canon, to "first do no harm." In addition, the avoidance of physical and psychological harm was established as a standard for research in the Nuremberg Code, 1946–49. At this time, the significant risks to the fetus and physical well being of a child created by somatic cell nuclear transplantation cloning outweigh arguably beneficial uses of the technique.

It is important to recognize that the technique that produced Dolly the sheep was successful in only 1 of 277 attempts. If attempted in humans, it would pose the risk of hormonal manipulation in the egg donor; multiple miscarriages in the birth mother; and possibly severe developmental abnormalities in any resulting child. Clearly the burden of proof to justify such an experimental and potentially dangerous technique falls on those who would carry out the experiment. Standard practice in biomedical science and clinical care would never allow the use of a medical drug or device on a human being on the basis of such a preliminary study and without much additional animal research. Moreover, when risks are taken with an innovative therapy, the justification lies in the prospect of treating an illness in a patient, whereas, here no patient is at risk until the innovation is employed. Thus, no conscientious physician or Institutional Review Board should approve attempts to use somatic cell nuclear transfer to create a child at this time. For these reasons, prohibitions are warranted on all attempts to produce children through nuclear transfer from a somatic cell at this time.

Even on this point, however, NBAC has noted some difference of opinion. Some argue, for example, that prospective parents are already allowed to conceive, or to carry a conception to term, when there is a significant risk—or even certainty—that the child will suffer from a serious genetic disease. Even when others think such conduct is morally wrong, the parents' right to reproductive freedom takes precedence. Since many of the risks believed to be associated with somatic cell nuclear transfer may be no greater than those associated with genetic disorders, some contend that such cloning should be subject to no more restriction than other forms of reproduction (Brock, 1997).

And, as in any new and experimental clinical procedure, harms cannot be accurately determined until trials are conducted in humans. Law professor John Robertson noted before NBAC on March 13, 1997 that:

> [The] first transfer [into a uterus] of a human [embryo] clone [will occur] before we know whether it will succeed. . . . [Some have argued therefore] that the first transfers are somehow unethical . . . experimentation on the resulting child, because one does not know what is going to happen, and one is . . . possibly leading to a child who could be disabled and have developmental difficulties. . . . [But the] child who would result would not have existed but for the procedure at issue, and [if] the intent there is actually to benefit that child by bringing it into being . . . [this] should be classified as experimentation for [the child's] benefit and thus it would fall within recognized exceptions. . . . We have a very different set of rules for experimentation intended to benefit [the experimental subject] (Robertson, 1997).

But the argument that somatic cell nuclear transfer cloning experiments are "beneficial" to the resulting child rests on the notion that it is a "benefit" to be brought into the world as compared to being left unconceived and unborn. This metaphysical argument, in which one is forced to compare existence with non-existence, is problematic. Not only does it require us to compare something unknowable—non-existence—with something else, it also can lead to absurd conclusions if taken to its logical extreme. For example, it would support the argument that there is no degree of pain and suffering that cannot be inflicted on a child, provided that the alternative is never to have been conceived. Even the originator of this line of analysis rejects this conclusion.[2]

In addition, it is true that the actual risks of physical harm to the child born through somatic cell nuclear transfer cannot be known with certainty unless and until research is conducted on human beings. It is likewise true that if we insisted on absolute guarantees of no risk before we permitted any new medical intervention to be attempted in humans, this would severely hamper if not halt completely the introduction of new therapeutic interventions, including new methods of responding to infertility. The assertion that we should regard attempts at human cloning as "experimentation for [the child's] benefit" is not persuasive.

Cloning and Individuality

In addition to physical harms, many worry about psychological harms associated with such cloning. One of the forms of psychological harm most frequently mentioned is the possible loss of a sense of uniqueness.

Many argue that somatic cell nuclear transfer cloning creates serious issues of identity and individuality and forces us to reconsider how we define ourselves. In his testimony before NBAC March 13, 1997, Gilbert Meilaender commented on the importance of genetic uniqueness not only for individuals but in the eyes of their parents:

> Our children begin with a kind of genetic independence of us, their parents. They replicate neither their father nor their mother. That is a reminder of the independence that we must eventually grant to them and for which it is our duty to prepare them. To lose even in principle this sense of the child as gift will not be good for children (Meilaender, 1997).

The concept of creating a genetic twin, although separated in time, is one aspect of somatic cell nuclear transfer cloning that most find both troubling and fascinating. The phenomenon of identical twins has intrigued human cultures across the globe, and throughout history (Schwartz, 1996). It is easy to understand why identical twins hold such fascination. Common experience demonstrates how distinctly different twins are, both in personality and in personhood. At the same time, observers cannot help but imbue identical bodies with some expectation that identical persons occupy those bodies, since body and personality remain intertwined in human intuition. With the prospect of somatic cell nuclear transfer cloning comes a scientifically inaccurate but nonetheless instinctive fear of multitudes of identical bodies, each housing personalities that are somehow less than distinct, less unique, and less autonomous than usual.

Is there a moral or human right to a unique identity, and if so would it be violated by this manner of human cloning? For such somatic cell nuclear transfer cloning to violate a right to a unique identity, the relevant sense of identity would have to be genetic identity, that is a right to a unique unrepeated genome. Even with the same genes, two individuals—for example homozygous twins—are distinct and not identical, so what is intended must be the various properties and characteristics that make each individual qualitatively unique and different than others. Does having the same genome as another person undermine that unique qualitative identity?

Along these lines of inquiry some question whether reproduction using somatic cell nuclear transfer would violate what philosopher Hans Jonas called a right to ignorance, or what philosopher Joel Feinberg called a right to an open future, or what Martha Nussbaum called the quality of "separateness" (Jonas, 1974; Feinberg, 1980; Nussbaum, 1990). Jonas argued that human cloning, in which there is a substantial time gap between the beginning of the lives of the earlier and later twin, is fundamentally different from the simultaneous beginning of the lives of homozygous twins that occur in nature. Although contemporaneous twins begin their lives with the same genetic inheritance, they also begin their lives or biographies at the same time, in ignorance of what the twin who shares the same genome will by his or her choices make of his or her life. To whatever extent one's genome determines one's future, each life begins ignorant of what that determination will be, and so remains as free to choose a future as are individuals who do not have a twin. In this line of reasoning, ignorance of the effect of one's genome on one's future is necessary for the spontaneous, free, and authentic construction of a life and self.

A later twin created by cloning, Jonas argues, knows, or at least believes he or she knows, too much about him or herself. For there is already in the world another person, one's earlier twin, who from the same genetic starting point has made the life choices that are still in the later twin's future. It will seem that one's life has already been lived and played out by another, that one's fate is already determined, and so the later twin will lose the spontaneity of authentically creating and becoming his or her own self. One will lose the sense of human possibility in freely creating one's own future. It is tyrannical, Jonas claims, for the earlier twin to try to determine another's fate in this way.

And even if it is a mistake to believe such crude genetic determinism according to which one's genes determine one's fate, what is important for one's experience of freedom and ability to create a life for oneself is whether one thinks one's future is open and undetermined, and so still to be largely determined by one's own choices. One might try to interpret Jonas' objection so as not to assume either genetic determinism, or a belief in it. A later twin might grant that he or she is not destined to follow in his or her earlier twin's footsteps, but that nevertheless the earlier twin's life would always haunt the later twin, standing as an undue influence on the latter's life, and shaping it in ways to which others' lives are not vulnerable.

In a different context, and without applying it to human cloning, Feinberg has argued for a child's right to an open future. This requires that others raising a child not close off the future possibilities that the child would otherwise have by constructing his or her own life. One way this right to an open future would be violated is to deny even a basic education to a child, and another way might be to create the child as a later twin so that he or she will believe its future has already been set by the choices made and the life lived by the earlier twin.

On the other hand, all of these concerns are not only quite speculative, but are directly related to certain specific cultural values. Someone created through the use of somatic cell nuclear transfer techniques may or may not believe that their future is relatively constrained. Indeed, they may believe the opposite. In addition, quite normal parenting usually involves many constraints on a child's behavior that children may resent. Moreover, Feinberg's argument does not apply, if the belief is false and it can be shown to be false.

Thus, a central difficulty in evaluating the implications for somatic cell nuclear transfer cloning of a right either to ignorance or to an open future, is whether the right is violated merely because the later twin may be likely to believe that its future is already determined, even if that belief is clearly false and supported only by the crudest genetic determinism. Moreover, what such a twin is likely to believe will depend on the facts that emerge and what scientists and ethicists claim.

Cloning and the Family

Among those concerns that are not focused on arguments about harm to the child are a set of worries about use of such cloning as a means of control. There are concerns, for example, about possibly generating large numbers of people whose life choices are limited by their own constrained self-image or by the constraining expectations of others. From this image of less-than-autonomous children comes the fear, however misplaced, of technology creating armies of cloned soldiers, each diminished in his or her physical individuality and thereby diminished in their psychological autonomy. Similarly, this expectation of diminished autonomy underlies the eugenic arguments that have led many to speculate about the possibility of cloning "desirable" or "evil" people, ranging from actors to dictators of various stripes to distinguished religious leaders. Complicating matters even further, this misplaced belief in the ability of genes to fully determine behavior and personality amplifies the image, so that in the end one imagines being able to make either armies of complacent workers, crazed soldiers, brilliant musicians, or beatific saints.

Although such fears are based, as noted in Chapter Two, on gross misunderstandings of human biology and psychology, they are nonetheless fears that have been voiced. In addition, these same concerns also manifest themselves in fears that underlie the characterization of somatic cell nuclear transfer cloning as a form of "making" children rather than "begetting" children. With cloning, the total genetic blueprint of the cloned individual is selected and determined by the human artisans. This, according to Kass:

. . . would be taking a major step into making man himself simply another one of the man made things. Human nature becomes merely the last part of nature to succumb to the technological project which turns all of nature into raw material at human disposal. . . . As with any product of our making, no matter how excellent, the artificer stands above it, not as an equal but as a superior, transcending it by his will and creative prowess (Kass, 1997).

For many, this kind of relationship is inconsistent with an ideal of parenting, in which parents embrace not only the similarities between themselves and their children but also the differences, and in which they accept not only the developments they sought to bring about through care and teaching but also the serendipitous developments they never planned for or anticipated (Rothenberg, 1997).

Of course, parents already exercise great control over their offspring, through means as varied as contraception to control of the timing and spacing of births, to genetic screening and use of donor gametes to avoid genetic disorders, to organized medical and educational interventions to guide physical and intellectual development. These interventions exist along a spectrum of control over development. Somatic cell nuclear transfer cloning, some fear, offers the possibility of virtually complete control over one important aspect of a child's development, his or her genome, and it is the completeness of this control, even if only over this partial aspect of human development, that is alarming to many people and invokes images of manufacturing children according to specification. The lack of acceptance this implies for children who fail to develop according to expectations, and the dominance it introduces into the parent-child relationship, is viewed by many as fundamentally at odds with the acceptance, unconditional love, and openness characteristic of good parenting. Meilaender addressed both the mystery of reproduction and fears about it veering toward a means of production in his testimony before NBAC:

But whatever we say of [other reproductive technologies], surely human cloning would be a new and decisive turn on this road. Far more emphatically a kind of production. Far less a surrender to the mystery of the genetic lottery which is the mystery of the child who replicates neither Father nor Mother but incarnates their union. Far more an understanding of the child as a product of human will (Meilaender, 1997).

Questions are raised, as well, about the effect such interventions will have on a particular child. Will the child himself or herself feel less independent from the nucleus donor than a child ordinarily would from a parent? Will the knowledge of how one's genetic profile developed in another person at another time leave the child feeling that his character is as predetermined as his eye or hair color? Even if the child feels completely independent of the nucleus donor, will others regard the child as a copy or a successor to that donor? If so, will such expectations on the part of others warp the child's emerging self-understanding?

Finally, some critics of such cloning are concerned that the legal or social status of the child arising from nuclear transfer of somatic cells may be uncertain. For some, the disparity between the child's genetic and social identity threatens the stability of the family. Is the child who results from somatic cell nuclear transfer the sibling or the child of its parents? The child or the grandchild of its grandparents? From this perspective the child's psychological and social well-being may be in doubt or even endangered. Ambiguity over parental roles may undermine the child's sense of identity. It may be harder for a child to achieve independence from a parent who is also his or her twin.

At the same time, others are not persuaded by such objections. Children born through assisted reproductive technologies may also have complicated relationships to genetic, gestational, and rearing parents. Skeptics of this point of view note that there is no evidence that confusion over family roles has harmed children born through assisted reproductive technologies, although the subject has not been carefully studied.

Potential Harms to Important Social Values

Those with grave reservations about somatic cell nuclear transfer cloning ask us to imagine a world in which cloning human beings via somatic cell nuclear transfer were permitted and widely practiced. What kind of people, parents, and children would we become in such a world? Opponents fear that such cloning to create children may disrupt the interconnected web of social values, practices, and institutions that support the healthy growth of children. The use of such cloning techniques might encourage the undesirable attitude that children are to be valued according to how closely they meet parental expectations, rather than loved for their own sake. In this way of looking at families and parenting, certain values are at the heart of those relationships, values such as love, nurturing, loyalty, and steadfastness. In contrast, a world in which such cloning were widely practiced would give, the critics claim, implicit approval to vanity, narcissism, and avarice. To these critics, changes that undermine those deeply prized values should be avoided if possible. At a minimum, such undesirable changes should not be fostered by public policies.

On the other hand, others are not persuaded by these objections. First, many social observers point out that if strongly held moral values are in decline, there are likely many complex reasons for this, which would not be addressed by a ban on cloning in this fashion. Furthermore, skeptics argue that people can, and do, adapt in socially redeeming ways, to new technologies. In their view, a child born through somatic cell nuclear transfer could be loved and accepted like any other child, and not disrupt important family and kinship relations.

The strength of public reaction, however, reflects a deep concern that somehow many important social values could be harmed in a society where such cloning were widely used. In his testimony before the Commission on March 13, 1997, bioethicist Leon Kass summarized many of the widely held concerns regarding the possibility of cloning human beings via somatic cell nuclear transfer when he noted:

> Almost no one sees any compelling reason for human cloning. Almost everyone anticipates its possible misuses and abuses. Many feel oppressed by the sense that there is nothing we can do to prevent it from happening and this makes the prospect seem all the more revolting. Revulsion is surely not an argument. . . . But . . . in crucial cases repugnance is often the emotional bearer of deep wisdom beyond reason's power fully to articulate it (Kass, 1997).

But some people, however, argue against relying on moral intuition to set public policy. While it is certainly true that repugnance may be the bearer of wisdom, it may also be the bearer of simple and thoughtless prejudice. In her testimony before NBAC on March 14, 1997, bioethicist Ruth Macklin challenged the inclination to take as axiomatic the proposition that to be born as a result of using these techniques is to be harmed or at least to be wronged:

> Intuition has never been a reliable epistemological method, especially since people notoriously disagree in their moral intuitions. . . . If objectors to cloning can identify no greater harm than a supposed affront to the dignity of the human species, that is a flimsy basis on which to erect barriers to scientific research and its applications (Macklin, 1997).

Nevertheless, opponents assert that this new type of cloning tempts human beings to transgress moral boundaries and to grasp for powers that are properly outside human control. Ancient Greek literature and many Biblical interpretations emphasize that human beings occupy a moral position between other forms of life and the divine. In particular, humans should not consider themselves as omnipotent over nature. From this perspective, respecting limits is to respect the appropriate place of humankind in the universe and to ensure that technology is not allowed to push aside critical social and moral commitments. This view need not be tied to a single religious doctrine, a particular view of God, or even a belief in God. However, these objections are often expressed in religious terms. For example, critics talk of how the ability to create children through somatic cell nuclear transfer may tempt us to seek immortality, to usurp the role of God, or to violate divine commands.

On the other hand, some observers do not see this type of cloning as dramatically new or extreme, especially when compared to other assisted reproductive technologies. Robertson notes:

> In an important sense cloning is not the most radical thing on the horizon. Much more significant, I think, would be the ability to actually alter or manipulate the genome of offspring. Cloning takes a genome as it is . . . and might replicate it. . . . [T]hat is much less ominous than having an ability to take a given genome and either add or take out a gene which could then lead to a child being born with characteristics other than it would have had with the genome it started with (Robertson, 1997).

Finally, critics have also raised questions about an inappropriate use of scarce resources. The generation of children through somatic cell nuclear transfer would divert scarce resources, including the skills of researchers and clinicians, from more pressing social and medical needs. These considerations about allocation of resources are particularly pertinent if public funds would be involved. In the words of theologian Nancy Duff:

> When considering research into human cloning we must look at the responsible use of limited resources. . . . [I]t is mandatory to ask whether other research projects will serve a greater number of people than research on human cloning and take the answer to that seriously (Duff, testimony, 1997).

Treating People As Objects

Some opponents of somatic cell nuclear cloning fear that the resulting children will be treated as objects rather than as persons. This concern often underlies discussions of whether such cloning amounts to "making" rather than "begetting" children, or whether the child who is created in this manner will be viewed as less than a fully independent moral agent. In sum, will being cloned from the somatic cell of an existing person result in the child being regarded as less of a person whose humanity and dignity would not be fully respected.

One reason this discussion can be hard to capture and to articulate is that certain terms, such as "person," are used differently by different people.[3] What is common to these various views, however, is a shared understanding that being a "person" is different from being the manipulated "object" of other people's desires and expectations. Writes legal scholar Margaret Radin,

The person is a subject, a moral agent, autonomous and self-governing. An object is a non-person, not treated as a self-governing moral agent. . . . [By] "objectification of persons," we mean, roughly, "what Kant would not want us to do."[4]

That is, to objectify a person is to act towards the person without regard for his or her own desires or well-being, as a thing to be valued according to externally imposed standards, and to control the person rather than to engage her or him in a mutually respectful relationship. Objectification, quite simply, is treating the child as an object—a creature less deserving of respect for his or her moral agency. Commodification is sometimes distinguished from objectification and concerns treating persons as commodities, including treating them as a thing that can be exchanged, bought or sold in the marketplace. To those who view the intentional choice by another of one's genetic makeup as a form of manipulation by others, somatic cell nuclear transfer cloning represents a form of objectification or commodification of the child.

Some may deny that objectification is any more a danger in somatic cell nuclear transfer cloning than in current practices such as genetic screening or, in the future perhaps, gene therapy. These procedures aim either to avoid having a child with a particular condition, or to compensate for a genetic abnormality. But to the extent that the technology is used to benefit the child by, for example, allowing early preventive measures with phenylketonuria, no objectification of the child takes place.

When such cloning is undertaken not for any purported benefit of the child himself or herself, but rather to satisfy the vanity of the nucleus donor, or even to serve the need of someone else, such as a dying child in need of a bone marrow donor, then some would argue that it goes yet another step toward diminishing the personhood of the child created in this fashion. The final insult, opponents argue, would come if the child created through somatic cell nuclear transfer is regarded as somehow less than fully equal to the other human beings, due to his or her diminished physical uniqueness and the diminished mystery surrounding some aspects of his or her future, physical development.

Eugenic Concerns

The desire to improve on nature is as old as humankind. It has been played out in agriculture through the breeding of special strains of domesticated animals and plants. With the development of the field of genetics over the past 100 years came the hope that the selection of advantageous inherited characteristics—called eugenics, from the Greek *eugenes* meaning wellborn or noble in heredity—could be as beneficial to humankind as selective breeding in agriculture.

The transfer of directed breeding practices from plants and animals to human beings is inherently problematic, however. To begin, eugenic proposals require that several dubious and offensive assumptions be made. First, that most, if not all people would mold their reproductive behavior to the eugenic plan; in a country that values reproductive freedom, this outcome would be unlikely absent compulsion. Second, that means exist for deciding which human traits and characteristics would be favored, an enterprise that rests on notions of selective human superiority that have long been linked with racist ideology.

Equally important, the whole enterprise of "improving" humankind by eugenic programs oversimplifies the role of genes in determining human traits and characteristics. Little is known about correlation between genes and the sorts of complex, behavioral characteristics that are associated with successful and rewarding human lives; moreover, what little is known indicates that most such characteristics result from complicated interactions among a number of genes and the environment. While cows can be bred to produce more milk and sheep to have softer fleece, the idea of breeding humans to be superior would belong in the realm of science fiction even if one could conceive how to establish the metric of superiority, something that turns not only on the values and prejudices of those who construct the metric but also on the sort of a world they predict these specially bred persons would face.

Nonetheless, at the beginning of this century eugenic ideas were championed by scientific and political leaders and were very popular with the American public. It was not until they were practiced in such a grotesque fashion in Nazi Germany that their danger became apparent. Despite this sordid history and the very real limitations in what genetic selection could be expected to yield, the lure of "improvement" remains very real in the minds of some people. In some ways, creating people through somatic cell nuclear transfer offers eugenicists a much more powerful tool than any before. In selective breeding programs, such as the "germinal choice" method urged by the geneticist H. J. Muller a generation ago (Kevles, 1995), the outcome depended on the usual "genetic lottery" that occurs each time a sperm fertilizes an egg, fusing their individual genetic heritages into a new individual. Cloning, by contrast, would allow the selection of a desired genetic prototype which would be replicated in each of the "offspring," at least on the level of the genetic material in the cell nucleus.

It might be enough to object to the institution of a program of human eugenic cloning—even a voluntary program—that it would rest on false scientific premises and hence be wasteful and misguided. But that argument might not be sufficient to deter those people who want to push the genetic traits of a population in a particular direction. While acknowledging that a particular set of genes can be expressed in variety of ways and therefore that cloning (or any other form of eugenic selection) does not guarantee a particular phenotypic manifestation of the genes, they might still argue that certain genes provide a better starting point for the next generation than other genes.

The answer to any who would propose to exploit the science of cloning in this way is that the moral problems with a program of human eugenics go far beyond practical objections of infeasibility. Some objections are those that have already been discussed in connection with the possible desire of individuals to use somatic cell nuclear transfer that the creation of a child under such circumstances could result in the child being objectified, could seriously undermine the value that ought to attach to each individual as an end in themselves, and could foster inappropriate efforts to control the course of the child's life according to expectations based on the life of the person who was cloned.

In addition to such objections are those that arise specifically because what is at issue in eugenics is more than just an individual act, it is a collective program. Individual acts may be undertaken for singular and often unknown or even unknowable reasons, whereas a eugenics program would propagate dogma about the sorts of people who are desirable and those who are dispensable. That is a path that humanity has tread before, to its everlasting shame. And it is a path to whose return the science of cloning should never be allowed to give even the slightest support. . . .

RECOMMENDATIONS OF THE COMMISSION

With the announcement that an apparently quite normal sheep had been born in Scotland as a result of somatic cell nuclear transfer cloning came the realization that, as a society, we must yet again collectively decide whether and how to use what appeared to be a dramatic new technological power. The promise and the peril of this scientific advance was noted immediately around the world, but the prospects of creating human beings through this technique mainly elicited widespread resistance and/or concern. Despite this reaction, the scientific significance of the accomplishment, in terms of improved understanding of cell development and cell differentiation, should not be lost. The challenge to public policy is to support the myriad beneficial applications of this new technology, while simultaneously guarding against its more questionable uses.

Much of the negative reaction to the potential application of such cloning in humans can be attributed to fears about harms to the children who may result, particularly psychological harms associated with a possibly diminished sense of individuality and personal autonomy. Others express concern about a degradation in the quality of parenting and family life. And virtually all people agree that the current risks of physical harm to children associated with somatic cell nuclear transplantation cloning justify a prohibition at this time on such experimentation.

In addition to concerns about specific harms to children, people have frequently expressed fears that a widespread practice of somatic cell nuclear transfer cloning would undermine important social values by opening the door to a form of eugenics or by tempting some to manipulate others as if they were objects instead of persons. Arrayed against these concerns are other important social values, such as protecting personal choice, particularly in matters pertaining to procreation and child rearing, maintaining privacy and the freedom of scientific inquiry, and encouraging the possible development of new biomedical breakthroughs.

As somatic cell nuclear transfer cloning could represent a means of human reproduction for some people, limitations on that choice must be made only when the societal benefits of prohibition clearly outweigh the value of maintaining the private nature of such highly personal decisions. Especially in light of some arguably compelling cases for attempting to clone a human being using somatic cell nuclear transfer, the ethics of policy making must strike a balance between the values society wishes to reflect and issues of privacy and the freedom of individual choice.

To arrive at its recommendations concerning the use of somatic cell nuclear transfer techniques, NBAC also examined long-standing religious traditions that often influence and guide citizens' responses to new technologies. Religious positions on human cloning are pluralistic in their premises, modes of argument, and conclusions. Nevertheless, several major themes are prominent in Jewish, Roman Catholic, Protestant, and Islamic positions, including responsible human dominion over nature, human dignity and destiny, procreation, and family life. Some religious thinkers argue that the use of somatic cell nuclear transfer cloning to create a child would be intrinsically immoral and thus could never be morally justified; they usually propose a ban on such human cloning. Other religious thinkers contend that human cloning to create a child could be morally justified under some circumstances but hold that it should be strictly regulated in order to prevent abuses.

The public policies recommended with respect to the creation of a child using somatic cell nuclear transfer reflect the Commission's best judgments about both the ethics of attempting such an experiment and our view of traditions regarding limitations on individual actions in the name of the common good. At present, the use of this technique to create a child would be a premature experiment that exposes the developing child to unacceptable risks. This in itself is sufficient to justify a prohibition on cloning human beings at this time, even if such efforts were to be characterized as the exercise of a fundamental right to attempt to procreate. More speculative psychological harms to the child, and effects on the moral, religious, and cultural values of society may be enough to justify continued prohibitions in the future, but more time is needed for discussion and evaluation of these concerns.

Beyond the issue of the safety of the procedure, however, NBAC found that concerns relating to the potential psychological harms to children and effects on the moral, religious, and cultural values of society merited further reflection and deliberation. Whether upon such further deliberation our nation will conclude that the use of cloning techniques to create children should be allowed or permanently banned is, for the moment, an open question. Time is an ally in this regard, allowing for the accrual of further data from animal experimentation, enabling an assessment of the prospective safety and efficacy of the procedure in humans, as well as granting a period of fuller national debate on ethical and social concerns. The Commission therefore concluded that there should be imposed a period of time in which no attempt is made to create a child using somatic cell nuclear transfer.

Within this overall framework the Commission came to the following conclusions and recommendations.

I.   The Commission concludes that at this time it is morally unacceptable for anyone in the public or private sector, whether in a research or clinical setting, to attempt to create a child using somatic cell nuclear transfer cloning. We have reached a consensus on this point because current scientific information indicates that this technique is not safe to use in humans at this time. Indeed, we believe it would violate important ethical obligations were clinicians or researchers to attempt to create a child using these particular technologies, which are likely to involve unacceptable risks to the fetus and/or potential child. Moreover, in addition to safety concerns, many other serious ethical concerns have been identified, which require much more widespread and careful public deliberation before this technology may be used.

The Commission, therefore, recommends the following for immediate action:

- A continuation of the current moratorium on the use of federal funding in support of any attempt to create a child by somatic cell nuclear transfer.
- An immediate request to all firms, clinicians, investigators, and professional societies in the private and non-federally funded sectors to comply voluntarily with the intent of the federal moratorium. Professional and scientific societies should make clear that any attempt to create a

child by somatic cell nuclear transfer and implantation into a woman's body would at this time be an irresponsible, unethical, and unprofessional act.

II.   The Commission further recommends that:
   •   Federal legislation should be enacted to prohibit anyone from attempting, whether in a research or clinical setting, to create a child through somatic cell nuclear transfer cloning. It is critical, however, that such legislation include a sunset clause to ensure that Congress will review the issue after a specified time period (three to five years) in order to decide whether the prohibition continues to be needed. If state legislation is enacted, it should also contain such a sunset provision. Any such legislation or associated regulation also ought to require that at some point prior to the expiration of the sunset period, an appropriate oversight body will evaluate and report on the current status of somatic cell nuclear transfer technology and on the ethical and social issues that its potential use to create human beings would raise in light of public understandings at that time.

III. The Commission also concludes that:
   •   Any regulatory or legislative actions undertaken to effect the foregoing prohibition on creating a child by somatic cell nuclear transfer should be carefully written so as not to interfere with other important areas of scientific research. In particular, no new regulations are required regarding the cloning of human DNA sequences and cell lines, since neither activity raises the scientific and ethical issues that arise from the attempt to create children through somatic cell nuclear transfer, and these fields of research have already provided important scientific and biomedical advances. Likewise, research on cloning animals by somatic cell nuclear transfer does not raise the issues implicated in attempting to use this technique for human cloning, and its continuation should only be subject to existing regulations regarding the humane use of animals and review by institution-based animal protection committees.
   •   If a legislative ban is not enacted, or if a legislative ban is ever lifted, clinical use of somatic cell nuclear transfer techniques to create a child should be preceded by research trials that are governed by the twin protections of independent review and informed consent, consistent with existing norms of human subjects protection.
   •   The United States Government should cooperate with other nations and international organizations to enforce any common aspects of their respective policies on the cloning of human beings.

IV. The Commission also concludes that different ethical and religious perspectives and traditions are divided on many of the important moral issues that surround any attempt to create a child using somatic cell nuclear transfer techniques. Therefore, we recommend that:

- The federal government, and all interested and concerned parties, encourage widespread and continuing deliberation on these issues in order to further our understanding of the ethical and social implications of this technology and to enable society to produce appropriate long-term policies regarding this technology should the time come when present concerns about safety have been addressed.

V. Finally, because scientific knowledge is essential for all citizens to participate in a full and informed fashion in the governance of our complex society, the Commission recommends that:

- Federal departments and agencies concerned with science should cooperate in seeking out and supporting opportunities to provide information and education to the public in the area of genetics, and on other developments in the biomedical sciences, especially where these affect important cultural practices, values, and beliefs.

---

1. In support of its analysis, NBAC commissioned a paper written by Dan Brock, Brown University, titled "Cloning Human Beings: An Assessment of the Ethical Issues Pro and Con." Some of the material in this chapter is derived from that paper.

2. There is one argument that has been used by several commentators to undermine the apparent significance of potential harms to a child created through somatic cell nuclear transfer (Chadwick 1982; Robertson 1994, 1997; Macklin 1994). The point derives from a general problem, called the non-identity problem, posed by the philosopher Derek Parfit and not originally directed to human cloning (Parfit 1984). This view argues that all the problems of having been born via such cloning are not net harms to the resulting child because they are not worse than no life at all. Parfit does not accept the above argument as sound. Instead, he believes that if one could have a different child without these burdens (for example, by using a different method of reproduction) there is as strong a moral reason to do so (Brock 1995).

3. Moral philosophers think about personhood when they construct and deploy their views of human choice and moral agency. For Kantians, personhood is about free will and reason. From the point of view of Kantian moral personality, all of us are identical as persons. Philosophers of mind think about personhood when they try to figure out what constitutes personal identity. For many of these philosophers, personal identity means having a continuous life story that incorporates a past and a

future for oneself. From the point of view of personal identity, all of us are different, unique, as persons. Psychoanalysts think about personhood when they relate the constants of human life and development to broad personality structures. From the psychoanalytic point of view, each of us manifests the same dynamic personality structures, yet no two of us do so in exactly the same way; we are all the same and also all different. Welfare rights activists and human rights activists may think about personhood: what is the minimum of necessary resources for a full human life? Some medical ethicists think about personhood while trying to decide at what point does life cease to be a human life worth living? Political theorists at times think about personhood in the context of trying to understand what are the basics of individuality that the state should recognize or underwrite? Parents think about personhood: what part do I play in making possible the fullest kind of human-ness for my children?" (Radin, 1995).

4. "Kantian ethical thought," writes Radin, "distinguishes morally between persons and objects. Rational beings possessing free will (persons) are autonomous; the moral law requires that persons be treated as ends, not means. Objects in the natural world that are not rational beings possessing free will are not persons, and may appropriately be used as means by persons. Kant's view requires that persons, moral agents, not be treated as objects, manipulated at the will of persons. Kant presented his basic principles of ethics in *Immanuel Kant, Groundwork of the Metaphysics of Morals* (1785), translated by H. J. Paton in *The Moral Law* (1948)." [Margaret Radin, "Reflections on Objectification," 65 *Southern California Law Review* 341 (November 1991), at footnote 4]

REFERENCES

Annas, G. J. "Regulatory models for human embryo cloning: The free market, professional guidelines, and government restrictions," *Kennedy Institute of Ethics Journal* (4)3:235–249, 1994.

Brock, D. "Cloning Human Beings: An Assessment of the Ethical Issues Pro and Con," paper prepared for NBAC, 1997.

Brock, D. W. "The non-identity problem and genetic harm," *Bioethics* 9:269–275, 1995.

Cahill, L. Testimony presented to the National Bioethics Advisory Commission, March 13, 1997.

Chadwick, R. F. "Cloning," *Philosophy* 57:201–209, 1982.

Coleman, "Playing God or playing scientist: A constitutional analysis of laws banning embryological procedures," 27 *Pacific Law Journal* 1331, 1996.

Duff, N. "Theological Reflections on Human Cloning," Testimony presented to the National Bioethics Advisory Commission, March 13, 1997.

Etzioni, A. *The Moral Dimension* (New York: The Free Press, 1990).

Feinberg, J. "The child's right to an open future," in *Whose Child? Children's Rights,*

Parental Authority, and State Power, W. Aiken and H. LaFollette (eds.) (Totowa, NJ: Rowman & Littlefield, 1980).

Glendon, M. A. *Rights Talk* (New York: The Free Press, 1991).

Gutmann, A., and D. Thompson. *Democracy and Disagreement* (Cambridge, MA: Belknap Press, 1996).

Jonas, H. *Philosophical Essays: From Ancient Creed to Technological Man* (Englewood Cliffs, NJ: Prentice-Hall, 1974).

Kass, L. "Why We Should Ban the Cloning of Human Beings," Testimony presented to the National Bioethics Advisory Commission, March 13, 1997.

Kevles, D. J. *In the Name of Eugenics* (Cambridge, MA: Harvard University Press, 1995).

Macklin, R. "Why We Should Regulate—But Not Ban—the Cloning of Human Beings," Testimony presented to the National Bioethics Advisory Commission, March 14, 1997.

Macklin, R. "Splitting embryos on the slippery slope: Ethics and public policy," *Kennedy Institute of Ethics Journal* 4:209–226, 1994.

Meilaender, G. "Remarks on Human Cloning to the National Bioethics Advisory Commission," Testimony presented to the National Bioethics Advisory Commission, March 13, 1997.

Mill, J. S. *On Liberty* (Indianapolis, IN: Bobbs-Merrill Publishing, 1859).

National Institutes of Health. *Report of the Human Embryo Research Panel* (Bethesda, MD: National Institutes of Health, 1994).

Nussbaum, M. C. "Aristotelian social democracy," in *Liberalism and the Good* 203, R. Bruce Douglass, et al. (eds.), pp. 217–226, 1990.

Parfit, D. *Reasons and Persons* (Oxford: Oxford University Press, 1984).

Posner, R. *Sex and Reason* (Cambridge, MA: Harvard University Press, 1992).

Radin, M. "Reflections on Objectification," 65 *Southern California Law Review* 341 (November 1991).

Radin, M. "The Colin Ruagh Thomas O'Fallon Memorial Lecture on Personhood," 74 *Oregon Law Review* 423 (Summer 1995).

Rhodes, R. "Clones, harms, and rights," *Cambridge Quarterly of Healthcare Ethics* 4:285–290, 1995.

Robertson, J. A. "A Ban on Cloning and Cloning Research is Unjustified," Testimony Presented to the National Bioethics Advisory Commission, March 14, 1997.

Robertson, J. A. "The question of human cloning," *Hastings Center Report* 24:6–14, 1994.

Robertson. "The scientist's right to research: A constitutional analysis," 51 *Southern California Law Review* 1203, 1977.

Rothenberg, K. Testimony before the Senate Committee on Labor and Human Resources, March 12, 1997.

Schwartz, H. *The Culture of Copy* (New York: Zone Books, 1996).

WHOSE SELF IS IT, ANYWAY?

PHILIP KITCHER

In the following essay, Philip Kitcher argues that it would be immoral for a couple to try to use cloning to create a child with specific qualities. By doing so, parents would be "demonstrating a crass failure to recognize their children as independent beings with the freedom to form their own sense of who they are and what their lives mean."

In a small range of cases, Kitcher admits that human cloning might be permissible to create a biological connection to a parent or parents. Such cases would be justified only when no other way exists to create such a connection, such as for lesbian parents. Overall, Kitcher does not think that human cloning is worth pursuing as sound social policy because, he implies, it would not contribute to a more just society.

Philip Kitcher has taught at the University of Vermont and University of Minnesota. In 1993–94, he was a senior fellow at the Library of Congress, where he reported on the social implications of the Human Genome Project. He is presently the Presidential Professor of Philosophy at the University of California at San Diego and recently published The Lives to Come: The Genetic Revolution and Human Possibilities.

In April 1988 Abe and Mary Ayala of Walnut, California, began living through every parent's nightmare: Anissa, their sixteen-year-old daughter, was diagnosed with leukemia. Without a bone-marrow transplant, Anissa would probably die within five years. But who could donate bone marrow that Anissa's immune system would not reject? Tests confirmed the worst: neither Abe, Mary nor their other child had compatible marrow.

The family embarked on a desperate plan. Abe, who had had a vasectomy years before, had it surgically reversed. Within months, at the age of forty-three, Mary became pregnant. The genetic odds were still three-to-one against a match between Anissa's bone marrow and that of the unborn child. The media got hold of the story, and the unbearable wait became a public agony.

Against all the odds a healthy daughter was born with compatible bone marrow. Fourteen months later, in June 1991, physicians extracted a few ounces of the child's marrow: the elixir that would save her older sister's life.

The story has a happy ending, but many people have found it at least slightly disturbing. Is it right for a couple to conceive one child to save another? Can someone brought into the world for such a well-defined purpose ever feel that she is loved for who she is? Thirty-seven percent of the people questioned in a contemporaneous *Time* magazine poll said they thought what the Ayalas had done was wrong; 47 percent believed it was justifiable.

Six years have passed and now a different, yet related, event a continent away has shaken the public's moral compass. Lamb number 6LL3, better known as Dolly, took the world by surprise last February when she was introduced as the first creature ever cloned from an adult mammal. Recognizing that what is possible with sheep today will probably be feasible with human beings tomorrow, commentators speculated about the legitimacy of cloning a Pavarotti or an Einstein, about the chances that a demented dictator might produce an army of supersoldiers, about the future of basketball in a world where a team of Larry Birds could play against a team of Michael Jordans. Polls showed that Mother Teresa was the most popular choice for person-to-be-cloned, but the film star Michelle Pfeiffer was not far behind, and Bill and Hillary Clinton, though tainted by controversy over alleged abuses of presidential power, also garnered some support.

Beyond all the fanciful talk, Dolly's debut introduces real and pressing moral issues. Cloning will not enable anyone to duplicate people like so many cookie-cutter gingerbread men, but it will pave the way for creating children who can fulfill their parents' preordained intentions. Families in the Ayalas' circumstances, for instance, would have a new option: Clone their dying child to give birth to another whose identical genetic makeup would guarantee them a compatible organ or a tissue match. Should they be allowed to exercise that option? The ethical implications of cloning balance on a fine line.

Society can probably blame Mary Wollstonecraft Shelley and her fervent imagination for much of the brouhaha over cloning. The Frankenstein story colors popular reception of the recent news, fomenting a potent brew of associations: many people assume that human lives can be made to order, that there is something vaguely illicit about the process, and, of course, that it is all going to turn out disastrously. Reality is much more complicated—and more sobering—so one should preface debates about the morality of human cloning with a clear understanding of the scientific facts.

As most newspaper readers know by now, the recent breakthroughs in cloning did not come from one of the major centers of the genetic revolution, but from the far less glamorous world of animal husbandry and agricultural research. A team of investigators at the Roslin Institute, near Edinburgh, Scotland, led by Ian Wilmut, conjectured that past efforts to clone mammals had failed because the cell that supplied the nucleus and the egg that received it were at different stages of the cell cycle. Applying well-known techiques from cell biology, Wilmut "starved" the cells so that both were in an inactive phase at the time of transfer. Inserting nuclei from adult sheep cells in that quiescent phase gave rise to a number of embryos, which were then implanted into ewes. In spite of a high rate of miscarriage, one of the pregnancies continued to term. After beginning with 277 transferred adult nuclei, Wilmut and his coworkers obtained one healthy lamb: the celebrated Dolly.

Wilmut's achievement raises three important questions about the prospect of human cloning: Will it be possible to undertake the same operations on human cells? Will cloners be able to reduce the high rate of failure? And just what is the relation between a clone obtained through nuclear transplantation and the animals, born in the usual way, from which the clone is derived?

Answers to the first two questions are necessarily tentative; predicting even the immediate trajectory of biological research is always vulnerable to contingencies. In the late 1960s, for example, after the developmental biologist J. B. Gurdon, now of the University of Cambridge, produced an adult frog through cloning, it seemed that cloning all kinds of animals was just around the corner; a few years later, the idea of cloning adult mammals had returned to the realm of science fiction. But leaving aside any definite time frame, one can reasonably expect that Wilmut's technique will eventually work on human cells and that failure rates will be reduced.

What about the third question, however, the relation between "parent" animal and clone? There one can be more confident. Dolly clearly has the same nuclear genetic material as the ewe that supplied the inserted nucleus. A second ewe supplied the egg into which that nucleus was inserted; hence Dolly's mitochondrial DNA came from another source. Indeed, though the exact roles played by mitochondrial DNA and other contents of the cytoplasm in vertebrate development are still unclear, one can say this much: Dolly's early development was shaped by the interaction between the DNA in the nucleus and the contents of the egg cytoplasm—the contributions of two adult females. A third sheep, the ewe into which the embryonic Dolly was implanted, provided Dolly with a uterine environment. Dolly thus has three mothers—nuclear mother, egg mother and womb mother—and no father (unless, of course, one accords that honor to Wilmut for his guiding role).

Now imagine Holly, a human counterpart of Dolly. You might think Holly would be similar to her nuclear mother, perhaps nearly identical, particularly if the mother of the nuclear mother were also the womb mother, and if either that woman or the nuclear mother were the egg mother. Such a hypothetical circumstance would ensure that Holly and her donor shared a similar gestation experience, as well as both nuclear and mitochondrial DNA. (Whether they would share other cytoplasmic constituents is anyone's guess, because the extent of the differences among eggs from a single donor is still unknown.)

But even if all Holly's genetic material and her intrauterine experience matched those of a single donor, Holly would not be an exact replica of that human being. Personal identity, as philosophers since John Locke have recognized, depends as much on life experiences as on genetics. Memories, attitudes, prejudices and emotional attachments all contribute to the making of a person. Cloning creates babies, not fully formed adults, and babies mature through a series of unique events. You could not hope to ensure the survival of your individual consciousness by arranging for one of your cells to be cloned. Megalomaniacs with intimations of immortality need not apply.

Other environmental factors would also lead to differences between Holly and her donor. For one thing, the two would likely belong to different generations, and the gap in their ages would correspond to changes in educational trends, the adolescent subculture and other aspects of society that affect children's development. Perhaps even more important, Holly and her donor would be raised in different families, with different friends, close relatives, teachers, neighbors and mentors. Even if the same couple acted as parents to both, the time gap would change the familial circumstances.

Identical twins reared together are obviously similar in many respects, but even they are by no means interchangeable; for instance, 50 percent of male identical twins who are gay have a twin who is not. Small differences in shared environments clearly play a large role. How much more dissimilarity, then, can be anticipated, given the much more dramatic variations that would exist between clones and their donors?

There will never be another you. If you hoped to fashion a son or daughter exactly in your image, you would be doomed to disappointment. Nevertheless, you might hope to take advantage of cloning technology to have a child of a certain kind—after all, the most obvious near-term applications for cloning lie in agriculture, where the technique could be used to perpetuate certain useful features of domestic animals, such as their capacity for producing milk, through succeeding generations. Some human characteristics are directly linked to specific genes and are therefore more amenable to manipulation—eye color, for instance. But in cloning, as in a good mystery novel, nothing is quite as simple as it seems.

Imagine a couple who are determined to do what they can to create a Hollywood star. Fascinated by the color of Elizabeth Taylor's eyes, they obtain a tissue sample from the actress and clone a young Liz. Will they succeed in creating a girl who possesses exact copies of the actress's celebrated eyes? Probably not. Small variations that occur at the cellular level during growth could modify the shape of the girl's eye sockets so that the eye color would no longer have its bewitching effect. Would the Liz clone still capture the hearts of millions? Perhaps the eyes would no longer have it.

Of course, Taylor's beauty and star appeal rest on much more than eye color. But the chances are that other physical attributes—height, figure, complexion, facial features—would also be somewhat different in a clone. Elizabeth II might overeat, for instance, or play strenuous sports, so that as a young adult her physique would be fatter or leaner than Elizabeth I's. Then there are the less tangible attributes that contribute to star quality: character and personal style. Consider what goes into something as apparently simple as a movie star's smile. Capturing as it does the interplay between physical features and personality, a smile is a trademark that draws on a host of factors, from jaw shape to sense of humor. How can anything so subtle ever be duplicated?

Fantasies about cloning Einstein, Mother Teresa or Yo-Yo Ma are equally doomed. The traits people value most come about through a complex interaction between genotypes and environments. By fixing the genotype one can only increase the chances—never provide a guarantee—of achieving one's desired results. The chances of artificially fashioning a person of true distinction in any area of complex human activity, whether it be science, philanthropy or artistic expression, are infinitesimal.

Although cloning cannot produce exact replicas or guarantee outstanding performance, it might be exploited to create a child who tends toward certain traits or talents. For example, had my wife and I wanted a son who would dominate the high school basketball court, we would have been ill-advised to reproduce in the old-fashioned way. At a combined height of just over eleven feet, we would have dramatically increased our chances by having a nucleus transferred from some strapping NBA star. And it is here, in the realm of the possible, that cloning scenarios devolve into moral squalor. By dabbling in genetic engineering, parents would be demonstrating a crass failure to recognize their children as independent beings with the freedom to form their own sense of who they are and what their lives mean.

Parents have already tried to shape and control their children, of course, even without the benefit of biological tools. The nineteenth-century English intellectual James Mill had a plan for his son's life, leading him to begin young John Stuart's instruction in Greek at age three and his Latin at age eight. John Stuart Mill's *Autobiography* is a quietly moving testament to the cramping effect of the life his eminent father had designed for him. In early adulthood, Mill *fils* suffered a nervous breakdown, from which he recovered, going on to a career of great intellectual distinction. But though John Stuart partly fulfilled his father's aspirations for him, one of the most striking features of his philosophical work is his passionate defense of human freedom. In *On Liberty* he writes: "Mankind are greater gainers by suffering each other to live as seems good to themselves, than by compelling each to live as seems good to the rest."

If the cloning of human beings is undertaken in the hope of generating a particular kind of person, then cloning is morally repugnant. The repugnance arises not because cloning involves biological tinkering but because it interferes with human autonomy. To discover whether circumstances might exist in which cloning would be morally acceptable, one must ask whether the objectionable motive can be removed. Three scenarios come immediately to mind.

First is the case of the dying child: Imagine a couple in a predicament similar to that of the Ayalas, which I described at the beginning of this essay. The couple's only son is dying and needs a kidney transplant within ten years. Unfortunately, neither parent can donate a compatible organ, and it may not be possible to procure an appropriate one from the existing donor pool. If a brother were produced by cloning, one of his kidneys could be transplanted to save the life of the elder son.

Second, the case of the grieving widow: A woman's beloved husband has been killed in an automobile accident. As a result of the same crash, the couple's only daughter lies in a coma with irreversible brain damage. The widow, who can no longer bear children, wants to have the nuclear DNA from one of her daughter's cells inserted into an egg supplied by another woman, so that a clone of her child can be produced through surrogate motherhood.

Third, the case of the loving lesbians: A lesbian couple wishes to have a child. Because they would like the child to be biologically connected to each of them, they request that a cell nucleus from one of them be inserted into an egg from the other, and that the embryo be implanted in the uterus of the woman who donated the egg.

No blatant attempt is made in any of these scenarios to direct the child's life; indeed, in some cases like these cloning may turn out to be morally justified. Yet lingering concerns remain. In the first scenario, and to a lesser extent in the second, the disinterested bystander suspects that children are being subordinated to the special purposes or projects of adults. Turning from John Stuart Mill to another great figure in contemporary moral theory, Immanuel Kant, one can ask whether any of the scenarios can be reconciled with Kant's injunction to "treat humanity, whether in your own person or in the person of another, always at the same time as an end and never simply as a means."

Perhaps the parents in the case of the dying child have no desire to expand their family; for them the younger brother would be simply a means of saving the really important life. And even if the parental attitudes were less callous, concerns would remain. In real case histories in which parents have borne a child to save an older sibling, their motives have been much more complex; the Ayala family seems a happy one, and the younger sister is thriving. Ironically, though, in such circumstances the parents' love for the younger child may be manifested most clearly if the project goes awry and the older child dies. Otherwise, the clone—and perhaps the parents as well—will probably always wonder whether he is loved primarily for his usefulness.

Similarly, the grieving widow might be motivated solely by nostalgia for the happy past, so that the child produced by cloning would be valuable only because she was genetically close to the dead. If so, another person is being treated as a means of understandable, but morbid, ends.

The case of the loving lesbians is the purest of the three. The desire to have a child who is biologically related to both of them is one that our society recognizes, at least for heterosexual couples, as completely natural and justifiable. There is no question in this scenario of imposing a particular plan on the nascent life—simply the wish to have a child who is the expression of the couple's mutual love. That is the context in which human cloning would be most defensible.

In recent decades, medicine has enabled many couples to overcome reproductive problems and bear their own biological children. Techniques of assisted reproduction have become mainstream because of a general belief that infertile couples have been deprived of something valuable, and that manipulating human cells is a legitimate response to their frustrations.

But do we, the members of a moral community, know what makes biological connections between parents and offspring valuable? Can we as a society assess the genuine benefits to the general welfare brought about by techniques of assisted reproduction, and do we want to invest in extending those techniques even further? Artificial insemination or in vitro fertilization could help the grieving widow and the lesbian couple in my scenarios; in both cases cloning would create closer biological connection—but one should ask what makes that extra degree of relatedness worth striving for. As for the parents of the dying child, one can simply hope that the continuing growth of genetic knowledge will provide improved methods of transplantation. By the time human cloning is a real possibility, advances in immunology may enable patients to tolerate tissue from a broader range of sources.

Should human cloning be banned? For the moment, while biology and medicine remain ignorant of the potential risks—the miscarriages and malformed embryos that could result—a moratorium is surely justified. But what if future research on nonhuman mammals proves reassuring? Then, as I have suggested, cloning would be permissible in a small range of cases. Those cases must satisfy two conditions: First, there must be no effort to create a child with specific attributes. Second, there must be no other way to provide an appropriate biological connection between parent and child. As people reflect on the second condition, perhaps some will be moved to consider just how far medicine should go to help people have children "of their own." Many families have found great satisfaction in rearing adopted children. Although infertile couples sometimes suffer great distress, further investment in technologies such as cloning may not be the best way to bring them relief.

The public fascination with cloning reached all the way to the White House almost immediately after Wilmut's epochal announcement. President Clinton was quick first to refer the issue to his National Bioethics Advisory Commission and then to ban federal funding for research into human cloning. The response was panicky, reflexive and disappointing. In the words of the editors of *Nature:* "At a time when the science policy world is replete with technology foresight exercises, for a US president and other politicians only now to be requesting guidance about [the implications of cloning] is shaming."

But though society and its leadership are woefully unprepared to handle cloning with policies based on forethought, many people race ahead irresponsibly with fantasies and fears. Human cloning becomes a titillating topic of discussion, while policy makers ignore the pressing ethical issues of the moment. In a fit of moral myopia, the U.S. government moves to reject human cloning because of potential future ills, while it institutes policies that permit existing children to live without proper health care and that endanger children's access to food and shelter.

The respect for the autonomy of lives and the duty to do what one can to let children flourish in their own ways should extend beyond hypothetical discussions about cloning. However strongly one may feel about the plights of loving lesbians, grieving widows or even couples with dying children, deciding how cloning might legitimately be applied to their troubles is not the most urgent moral or political question, or the best use of financial resources. I would hope that the public debate about new developments in biotechnology would ultimately spur our society to be more vigilant about applying the moral principles that we espouse but so often disregard.

Making demands for social investment seems quixotic, particularly when funds for the poor in the United States are being slashed and when other affluent countries are having second thoughts about the responsibilities of societies toward their citizens. The patronizing adjectives, such as "idealistic" and "utopian," that conservatives bestow on liberal programs do nothing to undermine the legitimacy of the demands. What is truly shameful is not that the response to the possibilities of cloning came so late, nor that the response has been so confused, but that the affluent nations have been so reluctant to think through the implications of time-honored moral principles and to design a coherent use of the new genetic science, technology and information for human well-being.

SCIENTIFIC DISCOVERIES

AND CLONING:

CHALLENGES FOR

PUBLIC POLICY

GEORGE ANNAS

After the announcement of Dolly's cloning, law professor George Annas was an immediate and passionate critic of the possibility of human cloning. He continues to defend this position. The following is his prepared testimony before a U.S. Senate subcommittee that held hearings on human cloning a few weeks after the news of Dolly.

Annas has been a well-known figure in American bioethics for twenty years, and his remarks are frequently reported in the national media.

Annas argues that cloning humans would violate a barrier of natural kinds, would have no good reason to serve as moral justification, and would violate the wisdom of decades of literature in science fiction that has warned us about the dangers of human cloning.

More ambitiously, Annas would revamp all present quasi-regulation of human experimentation through local Institutional Review Boards and establish a new federal agency, the Human Experimentation Agency. He thinks that people who wanted to originate a child through cloning should have the onus of proof before examiners of this agency to prove the safety and desirability of being allowed to have a child this way.

George Annas teaches at Boston University, where he is professor and chair in the Department of Health Law and where he founded the Law, Medicine, and Ethics Program.

SENATOR FRIST, THANK YOU FOR THE OPPORTUNITY TO APPEAR BE-
fore your subcommittee to address some of the legal and ethical aspects surrounding the prospect of human cloning. I agree with President Clinton that
we must "resist the temptation to replicate ourselves" and that the use of federal funds for the cloning of human beings should be prohibited. On the other
hand, the contours of any broader ban on human cloning require, I believe,
sufficient clarity to permit at least some research on the cellular level. This
hearing provides an important opportunity to help explore and define just what
makes the prospect of human cloning so disturbing to most Americans, and
what steps the federal government can take to prevent the duplication of
human beings without preventing vital research from proceeding.

I will make three basic points this morning: (1) the negative reaction to the
prospect of human cloning by the scientific, industrial and public sectors is
correct because the cloning of a human would cross a boundary that represents
a difference in kind rather than in degree in human "reproduction"; (2) there
are no good or sufficient reasons to clone a human; and (3) the prospect of
cloning a human being provides an opportunity to establish a new regulatory
framework for novel and extreme human experiments.

1. THE CLONING OF A HUMAN WOULD CROSS A NATURAL BOUNDARY THAT REPRESENTS A DIFFERENCE IN KIND RATHER THAN DEGREE OF HUMAN "REPRODUCTION."

There are those who worry about threats to biodiversity by cloning animals,
and even potential harm to the animals themselves. But virtually all of the
reaction to the appearance of Dolly on the world stage has focused on the
potential use of the new cloning technology to replicate a human being. What
is so simultaneously fascinating and horrifying about this technology that produced this response? The answer is simple, if not always well-articulated: *replication of a human by cloning would radically alter the very definition of a
human being by producing the world's first human with a single genetic parent.*
Cloning a human is also viewed as uniquely disturbing because it is the manufacture of a person made to order, represents the potential loss of individuality,
and symbolizes the scientist's unrestrained quest for mastery over nature for
the sake of knowledge, power, and profits.

Human cloning has been on the public agenda before, and we should recognize the concerns that have been raised by both scientists and policy makers
over the past twenty-five years. In 1972, for example, the House Subcommittee
on Science, Research and Development of the Committee on Science and
Astronautics asked the Science Policy Research Division of the Library of Congress to do a study on the status of genetic engineering. Among other things,

that report dealt specifically with cloning and parthenogenesis as it could be applied to humans. Although the report concluded that the cloning of human beings by nuclear substitution "is not now possible," it concluded that cloning "might be considered an advanced type of genetic engineering" if combined with the introduction of highly desirable DNA to "achieve some ultimate objective of genetic engineering." The Report called for assessment and detailed knowledge, forethought and evaluation of the course of genetic developments, rather than "acceptance of the haphazard evolution of the techniques of genetic engineering [in the hope that] the issues will resolve themselves."

Six years later, in 1978, the Subcommittee on Health and the Environment of the House Committee on Interstate and Foreign Commerce held hearings on human cloning in response to the publication of David Rorvick's *The Cloning of a Man*. All of the scientists who testified assured the committee that the supposed account of the cloning of a human being was fictional, and that the techniques described in the book could not work. One scientist testified that he hoped that by showing that the report was false it would also become apparent that the issue of human cloning itself "is a false one, that the apprehensions people have about cloning of human beings are totally unfounded." The major point the scientists wanted to make, however, was that they didn't want any regulations that might affect their research. In the words of one, "There is no need for any form of regulatory legislation, and it could only in the long run have a harmful effect."

Congressional discussion of human cloning was interrupted by the birth of Baby Louise Brown, the world's first IVF baby, in 1978. The ability to conceive a child in a laboratory not only added a new way (in addition to artificial insemination) for humans to reproduce without sex, but also made it possible for the first time for a woman to gestate and give birth to a child to whom she had no genetic relationship. Since 1978, a child can have at least five parents: a genetic and rearing father, and a genetic, gestational, and rearing mother. We pride ourselves as having adapted to this brave new biological world, but in fact we have yet to develop reasonable and enforceable rules for even so elementary a question as who among these five possible parents the law should recognize as those with rights and obligations to the child. Many other problems, including embryo storage and disposition, posthumous use of gametes, and information available to the child also remain unresolved.

IVF represents a striking technological approach to infertility; nonetheless the child is still conceived by the union of an egg and sperm from two separate human beings of the opposite sex. Even though no change in the genetics and biology of embryo creation and growth is at stake in IVF, society continues to wrestle with fundamental issues involving this method of reproduction twenty years after its introduction. Viewing IVF as a precedent for human cloning misses the point. Over the past two decades many ethicists have been accused of "crying wolf" when new medical and scientific technologies have been introduced. This may have been the case in some instances, but not here. This change in kind in the fundamental way in which humans can "reproduce" represents such a challenge to human dignity and the potential devaluation of human life (even comparing the "original" to the "copy" in terms of which is to be more valued) that even the search for an analogy has come up empty handed.

Cloning is replication, not reproduction, and represents a difference in kind not in degree in the manner in which human beings reproduce. Thus, although the constitutional right not to reproduce would seem to apply with equal force to a right not to replicate, to the extent that there is a constitutional right to reproduce (if one is able to), it seems unlikely that existing privacy or liberty doctrine would extend this right to replication by cloning.

2. THERE ARE NO GOOD OR SUFFICIENT REASONS TO CLONE A HUMAN.

When the President's Bioethics Commission reported on genetic engineering in 1982 in their report entitled *Splicing Life,* human cloning rated only a short paragraph in a footnote. The paragraph concluded: "The technology to clone a human does not—and may never—exist. Moreover, the critical nongenetic influences on development make it difficult to imagine producing a human clone who would act or appear 'identical'." (p. 10) The NIH Human Embryo Research panel that reported on human embryo research in September 1994 also devoted only a single footnote to this type of cloning. "Popular notions of cloning derive from science fiction books and films that have more to do with cultural fantasies than actual scientific experiments." (at p. 39) Both of these expert panels were wrong to disregard lessons from our literary heritage on this topic, thereby attempting to sever science from its cultural context.

Literary treatments of cloning help inform us that applying this technology to humans is too dangerous to human life and values. The reporter who described Dr. Ian Wilmut as "Dolly's laboratory father" couldn't have conjured up images of Mary Shelley's *Frankenstein* better if he had tried. Frankenstein was also his creature's father/god; the creature telling him: "I ought to be thy Adam." Like Dolly, the "spark of life" was infused into the creature by an electric current. Unlike Dolly, the creature was created as a fully grown adult (not a cloning possibility, but what many Americans fantasize and fear), and wanted more than creaturehood: he wanted a mate of his "own kind" with whom to live, and reproduce. Frankenstein reluctantly agreed to manufacture such a mate if the creature agrees to leave humankind alone, but in the end, viciously destroyed the female creature-mate, concluding that he has no right to inflict the children of this pair, "a race of devils," upon "everlasting generations." Frankenstein ultimately recognized his responsibilities to humanity, and Shelley's great novel explores virtually all the noncommercial elements of today's cloning debate.

The naming of the world's first cloned mammal also has great significance. The sole survivor of 277 cloned embryos (or "fused couplets"), the clone could have been named after its sequence number in this group (e.g., C-137), but this would have only emphasized its character as a produced product. In stark contrast, the name Dolly (provided for the public and not used in the scientific report in *Nature*) suggests an individual, a human or at least a pet. Even at the manufactured level a "doll" is something that produces great joy in our children and is itself harmless. Victor Frankenstein, of course, never named his creature, thereby repudiating any parental responsibility. The creature himself evolved into a monster when it was rejected not only by Frankenstein, but by society as well. Naming the world's first mammal-clone Dolly is meant to distance her from the Frankenstein myth both by making her appear as something she is not, and by assuming parental obligations toward her.

Unlike Shelley's, Aldous Huxley's *Brave New World* future in which all humans are created by cloning through embryo splitting and conditioned to join one of five worker groups, was always unlikely. There are much more efficient ways of creating killers or terrorists (or even workers) than through cloning—physical and psychological conditioning can turn teenagers into terrorists in a matter of months, rather than waiting some eighteen to twenty years for the clones to grow up and be trained themselves. Cloning has no real military or paramilitary uses. Even Hitler's clone would himself likely be quite a different person because he would grow up in a radically altered world environment.

It has been suggested, however, that there might be good reasons to clone a human. Perhaps most compelling is cloning a dying child if this is what the grieving parents want. But this should not be permitted. Not only does this encourage the parents to produce one child in the image of another, it also encourages all of us to view children as interchangeable commodities, because cloning is so different from human reproduction. When a child is cloned, it is not the parents that are being replicated (or are "reproducing") but the child. No one should have such dominion over a child (even a dead or dying child) as to be permitted to use its genes to create the child's child. Humans have a basic right not to reproduce, and human reproduction (even replication) is not like reproducing farm animals, or even pets. Ethical human reproduction properly requires the voluntary participation of the genetic parents. Such voluntary participation is not possible for a young child. Related human rights and dignity would also prohibit using cloned children as organ sources for their father/mother original. Nor is there any "right to be cloned" that an adult might possess that is triggered by marriage to someone with whom the adult cannot reproduce.

Any attempt to clone a human being should also be prohibited by basic ethical principles that prohibit putting human subjects at significant risk without their informed consent. Dolly's birth was a one in 277 embryo chance. The birth of a human from cloning might be technologically possible, but we could only discover this by unethically subjecting the planned child to the risk of serious genetic or physical injury, and subjecting a planned child to this type of risk could literally never be justified. Because we will likely never be able to protect the human subject of cloning research from serious harm, the basic ethical rules of human experimentation prohibit us from ever using it on humans.

3. DEVELOPING A REGULATORY FRAMEWORK FOR HUMAN CLONING

What should we do to prevent Dolly technology from being used to manufacture duplicate humans? We have three basic models for scientific/medical policy-making in the U.S.: the market, professional standards, and legislation. We tend to worship the first, distrust the second, and disdain the third. Nonetheless, the prospect of human cloning requires more deliberation about social and moral issues than either the market or science can provide. The market has no morality, and if we believe important values including issues of human rights and human dignity are at stake, we cannot leave cloning decisions to the market. The Biotechnology Technology Industry Organization in the U.S. has already taken the commendable position that human cloning should be prohibited by law. Science often pretends to be value-free, but in fact follows its own imperatives, and either out of ignorance or self-interest assumes that others are making the policy decisions about whether or how to apply the fruits of their labors. We disdain government involvement in reproductive medicine. But cloning is different in kind, and only government has the authority to restrain science and technology until its social and moral implications are adequately examined.

We have a number of options. The first is for Congress to simply ban the use of human cloning. Cloning for replication can (and should) be confined to nonhuman life. We need not, however, prohibit all possible research at the cellular level. For example, to the extent that scientists can make a compelling case for use of cloning technology on the cellular level for research on processes such as cell differentiation and senescence, and so long as any and all attempts to implant a resulting embryo into a human or other animal, or to continue cell division beyond a 14-day period are prohibited, use of human cells for research could be permitted. Anyone proposing such research, however, should have the burden of proving that the research is vital, cannot be conducted any other way, and is unlikely to produce harm to society.

The prospect of human cloning also provides Congress with the opportunity to go beyond ad hoc bans on procedures and funding, and the periodic appointment of blue ribbon committees, and to establish a Human Experimentation Agency with both rule-making and adjudicatory authority in the area of human experimentation. Such an agency could both promulgate rules governing human research and review and approve or disapprove research proposals in areas such as human cloning which local IRBs are simply incapable of providing meaningful reviews. The President's Bioethics panel is important and useful as a forum for discussion and possible policy development. But we have had such panels before, and it is time to move beyond discussion to meaningful regulation in areas like cloning where there is already a societal consensus.

One of the most important procedural steps a federal Human Experimentation Agency should take is to put the burden of proof on those who propose to do extreme and novel experiments, such as cloning, that cross reorganized boundaries and call deeply held societal values into question. Thus, cloning proponents should have to prove that there is a compelling reason to approve research on it. I think the Canadian Royal Commission on New Reproductive Technologies quite properly concluded that both cloning and embryo splitting have "no foreseeable ethically acceptable application to the human situation" and therefore should not be done. We need an effective mechanism to ensure that it is not.

WRONGFUL LIFE,

FEDERALISM,

AND PROCREATIVE

LIBERTY:

A CRITIQUE OF THE

NBAC CLONING REPORT

JOHN ROBERTSON

John Robertson was one of the few public critics of a proposed federal ban on human cloning by the NBAC. For twenty years Robertson has written about procreative ethics. He is a fellow of the Hastings Center, a resource center in bioethics. His articles on surrogate mothers and involuntary euthanasia of defective newborns are classics in bioethics.

In this piece, Robertson argues that the dangers of human cloning to children do not outweigh the dangers of restricting human procreative freedom. Robertson also believes that the NBAC Report contradicts itself in places. Robertson's most passionate claim is that society should not make it a federal crime to originate children by cloning.

John Robertson holds the Vinson & Elkins Chair in Law at the University of Texas School of Law. His most recent book is Children of Choice: Freedom and the New Reproductive Technologies.

THE ANNOUNCEMENT ON FEBRUARY 23, 1997 OF THE BIRTH OF Dolly, the sheep cloned from the mammary glands of an adult sheep, was potentially a paradigm-shifting event in human reproduction. Politicians quickly followed scientists and ethicists in commenting on its significance. Within 24 hours of the announcement, President Clinton had called for a moratorium on federal funding of human cloning research and directed his newly created National Bioethics Advisory Commission (NBAC) to report on human cloning in 90 days.[1]

On June 9, 1997 the NBAC recommended to the President that a federal criminal ban on human cloning be enacted for a three to five year period.[2] The President promptly submitted legislation calling for a five year ban to Congress, where three similar bills were pending.[3] Other countries and international organizations have taken similar positions.[4]

In contrast to the media explosion that greeted Dolly's birth announcement, coverage of NBAC's cloning report has been noticeably subdued. The report was scarcely noticed in the electronic and print media. Apparently in the months since Dolly's birth, human cloning had receded as a threat, as it became clear that human cloning was not an immediate possibility, and if ever feasible, would not turn out multiple copies of persons as originally imagined.

The NBAC Report is a good place to begin sorting out the complex issues that human cloning would raise. Overall, it has done a very creditable job of advancing the ethical and policy debate. The Report has usefully brought together relevant scientific and religious information, described the key ethical issues and policy alternatives, and come up with a plausible, though ultimately unsatisfying, interim solution.

Given the rushed circumstances under which the Commission worked, this is no small accomplishment. Appointed in September, 1996, the group of 18 persons had not yet tackled any major issue when they were asked to quickly devise national policy on human cloning. At least half of its members were not versed in bioethics, while those who were, like most bioethicists elsewhere, had not carefully considered cloning before. The fact that they did as well as they did in these circumstances is a credit to their able chairman, and to the hard work of the members and staff.

At the same time, no one should think that the NBAC Report is a model for public policy reports. Its account of issues is more descriptive than analytic. When analysis occurs, it is often sketchy or inconsistent, and doesn't always support its own conclusions. Often the writing is weak, and begs for proofreading and editing.[5]

Far from settling the public policy issues, the NBAC Report is best viewed as a first stab at dealing with the complicated issues raised by human cloning and other genetic selection techniques. With the NBAC's limited staff and time in mind, my critique focuses on the ethical and policy arguments it made for a federal criminal ban on cloning "at this time."

THE STRUCTURE OF THE NBAC REPORT

The NBAC Report, *Cloning Human Beings,* contains 110 pages divided into six chapters. It begins with a 13 page introduction, and is followed by a 26 page chapter "The Science and Application of Cloning," a 24 page chapter "Religious Perspectives," and a 24 page chapter "Ethical Considerations." Following the ethics chapter, there is a 20 page chapter on "Legal and Policy Considerations," and then a four page chapter of recommendations.[6]

Like the other chapters, the ethics chapter is largely descriptive. It gives brief descriptions of objections to human cloning based on physical safety, individuality, effects on the family, potential harms to important social values, treating people as objects, and eugenics. It also acknowledges that considerations of personal privacy, procreative liberty, and scientific inquiry are implicated, though it gives them much shorter shrift. For example, after noting three "exceptional" cases in which the arguments for cloning appear to be strong, it still finds the use of somatic cell nuclear transfer cloning now—and possibly in the future as well—to be morally unacceptable and deserving of federal prohibition. It concludes the chapter with some general remarks about comparing harms to benefits in determining public policy for human cloning.

The NBAC's main policy goal is to identify public policy toward cloning now when it is still undeveloped and experimental, and not yet shown to be safe in humans. It draws a sharp line between human cloning research, for which it does not recommend a ban, and attempts to create children by cloning. With regard to the latter, it recommends that federal law make it a crime to attempt to bring about the birth of a child through adult somatic cell nuclear transfer cloning.[7]

The ethics chapter, together with a discussion of policy options at the end of the ensuing chapter, contains the Report's reasoning for why human cloning is presently unethical and should be prohibited by federal law. Its argument for such a ban hinges largely on the safety issue. I focus first on the report's account of physical safety as a ground now for prohibition of cloning, and then turn to the question of whether public policy should incorporate the report's ethical judgment into federal criminal law.

PHYSICAL SAFETY AND THE ETHICS OF HUMAN CLONING

The NBAC justified its main recommendation in favor of a time-limited federal ban on all human cloning on grounds of physical safety.[8] Its position deserves careful analysis not only because it underpins the Report's call for a present ban, but also because of the implications of its premises for other issues in cloning and genetic selection.

The report's emphasis on physical safety is somewhat surprising. Issues about physical safety arise with any new medical procedure, but they were not at the top of anyone's list of fears about human cloning. There is always a danger that physicians, who have an interest in developing and using innovative procedures, will mislead patients about the prospect of success. In some cases patients may demand procedures when the risks and likelihood of success are unknown. While both are possibilities with somatic cell nuclear transfer, in a more realistic scenario few couples and doctors would be willing to use this technique if there were a significant risk of physical damage to offspring. People might differ somewhat in their willingness to accept risks, but it is highly unlikely that therapeutic or clinical transfers of embryos created by cloning would occur without extensive animal and laboratory research that first established its safety and efficacy.[9]

The Report's argument for a three-to-five year ban on human cloning is based on "the virtually universal concern regarding the current safety of attempting to use this technique in human beings."[10] It noted that "even if there were a compelling case in favor of creating a child in this manner, it would have to yield to one fundamental principle of both medical ethics and political philosophy—the injunction . . . to 'first do no harm.' "[11] It concluded: "At this time the significant risks to the fetus and physical well-being of a child created by somatic cell nuclear transplantation cloning outweigh arguably beneficial uses of this technique."[12]

In support of this position, it noted that Dr. Ian Wilmut's team took 277 tries to produce one cloned sheep. If attempted in humans, "it would pose the risk of hormonal stimulation in the egg donor; multiple miscarriages in the birth mother; and possibly severe developmental abnormalities in any resulting child."[13] Standard practice in biomedical science and clinical care would place the burden of proof on proponents to justify use of "an experimental and potentially dangerous technique" in humans. The Report asserted:

> . . . no conscientious physician or institutional review board should approve attempts to use somatic cell nuclear transfer to create a child at this time. For these reasons, prohibitions are warranted on all attempts to produce children through nuclear cell transfer from a somatic cell at this time.[14]

Although few persons would favor or would want to use cloning if children turned out to be damaged, the question of whether it would be unethical to do so, as well as the question of whether it would be unethical if many eggs were wasted, embryos were destroyed, or fetuses aborted, is much more complicated than at first appears. The Report mentions risks to fetuses as well as to children.[15] However, the welfare of children is clearly the greatest concern, and therefore will be the focus here.

The concern about harm to children is more problematic than the Report recognizes because the underlying principle on which it rests would make it unethical knowingly or intentionally to give birth to children who are not fully healthy, physically or psychologically. Such a principle is difficult to defend. Since it is not followed in most other reproductive situations, it is unclear why human cloning should be treated differently.

For example, most courts and commentators have objected to the idea of holding women criminally responsible for prenatal behavior that causes postnatal harm to offspring, even when the child would otherwise have been born healthy if the woman or man had not engaged in the behavior at issue.[16] Indeed, some persons object to any moral condemnation of pregnant women, even when they have knowingly or recklessly engaged in behavior that will cause a child who would otherwise have been born healthy to be born with mental or physical impairments.

Similarly, people are loath to sanction couples who knowingly bring genetically handicapped children into the world. Thus we do not normally think that couples who are carriers for Tay Sachs, sickle cell, or cystic fibrosis are acting wrongfully if they have not been screened for their carrier status, if they know they are carriers and yet intentionally conceive, or if having conceived, they then refuse prenatal diagnosis and abortion. Yet their conduct will cause a child to be born with a severe, physically debilitating disease. We also have come to accept a wide menu of assisted reproductive techniques that some people think are not good, at least psychologically, for resulting children. In all these cases the child would not have been born but for the conduct in question. Yet with a few technical exceptions to ensure that negligent professionals bear the extra rearing costs which their behavior causes, no court recognizes a claim of wrongful life on behalf of the child in those cases.[17]

It is hard to see what makes the case of cloning different. Here we must distinguish the early use of cloning by persons who have no interest in a family project and those who do. If a couple is willing to take the risk that embryos won't form or cleave, that they won't implant, that there will be a high rate of miscarriage, that the child will be born with some defect, and that they will then rear the child, it is hard to see why this is any worse than the other practices that could lead to physically-damaged offspring.

The Report's position is especially troubling because of its crude handling of the wrongful life problem, on which its view of the physical and psychological harm from cloning rests. The Report's claim that resulting children are harmed by cloning is open to challenge because but for the procedure in question, the child would not have been born. The difficulty of the claim is all the greater because there is no evidence that feared harms of cloning would cause such physical or psychological suffering that the child's very existence would be a wrong to that child.

The Report assumes that being born less than healthy or whole is always a harm to the child who results from nuclear transfer cloning, even when it has no alternative way to be born. It rejects the claim that the cloned child is not harmed when this is the only way it has to be born:

> But the argument that somatic cell nuclear transfer cloning experiments are 'beneficial' to the resulting child rest on the notion that it is a benefit to be brought into the world as compared to being left unconceived and unborn. This metaphysical argument, in which one is forced to compare existence and non-existence, is problematic. Not only does it require us to compare something unknowable—non-existence—with something else, it also can lead to absurd results if taken to its logical extreme. For example, it would support the argument that there is no degree of pain and suffering that cannot be inflicted on a child, provided that the alternative is never to have been conceived.[18]

There are several mistakes here. First, the claim that there is no harm to a cloned child from the fact of cloning because it had no other way to come into existence is not "metaphysical" based on the interests of nonexisting, potential children. The claim is made instead from the standpoint of the cloned child who now exists, not from the standpoint of someone waiting to exist.[19] It asks whether its present life is so full of suffering that it would now prefer nonexistence. This is comparing a present state of existence with non-existence as it appears from the perspective of the now existing person. True, we do not know what nonexistence is, but we do know that it is the absence of life and experience. This is a judgment that we allow people in terminal or debilitating conditions to make in refusing necessary medical treatment. There is no reason that we cannot apply it to existing persons whose very lives are said to harm them because they are the product of cloning or other controversial procedures.

Nor is the Report correct about the logical extreme to which this reasoning would lead. Neither cloning (or other techniques that are controversial because of alleged harm to offspring) are likely to impose such a high degree of physical harm. If they did, few couples would be interested, because would-be parents are generally interested in having a healthy child. If the technique at issue did in fact produce such enormous suffering that from the born child's own perspective nonexistence would have been preferable, then continuing the child's life would indeed be wrongful.[20] But that extreme state is highly unlikely to occur. Its possibility hardly provides a basis for finding harm in situations that amount to much less than a wrongful life which the child could not avoid if it is to live at all.

A third mistake is inconsistency. In calling for a ban on human cloning on grounds of the risk of physical safety to resulting offspring, the report relies on the very assumptions which it criticizes. If bringing the child into being is to harm it, even if there is no alternative way to be born, then according to NBAC, it is preferable that that child not exist at all. The NBAC is thus explicitly comparing existence and nonexistence, while at the same time asserting that they cannot be compared.[21]

In citing concerns about physical safety as a ground for finding cloning now to be unethical and deserving of prohibition, the Report has thus ignored the very real problems posed by attempts to protect children by preventing their birth altogether. If the ethical premise underlaying the physical safety argument is taken seriously, it would mean that any parental choice that would cause a child to be born less than healthy would unethically harm the child and deserve prohibition. Indeed, such an assumption underlays the Report's discussion of individuality, autonomy, and objectification as other possible reasons for banning cloning. All rely on the existence of states or conditions in resulting children when the child could not have existed but with the condition of concern.

Now few parents will be interested in assisted reproduction, cloning, or other genetic selection techniques if serious physical or psychological problems ensue. Is it nevertheless unethical or morally wrong to run risks that such effects might occur? If it is unethical, it will be necessary to find a basis for such a judgment other than harm to the resulting child, for the child in question has no other way to be born, and its very existence is unlikely to be so full of suffering as to be wrongful. Either the conclusion that it is unethical because of harm to children should be rejected, or some other basis for the ethical claim established. Accepting the premise that the child is harmed by existence itself, when its life cannot be shown to be so full of suffering as to amount to a wrongful life, would condemn a wide range of presently accepted reproductive and genetic selection behavior. Our unwillingness to draw this conclusion in other circumstances is further proof that there is something wrong with the report's assumptions.[22]

SHOULD FEDERAL LAW NOW CRIMINALIZE CLONING?

Even if one accepted the report's premise that children born after somatic cell nuclear transfer are harmed by their very existence, it would not follow that the alleged unethical behavior involved with bringing them into the world should be a federal crime. There are many forms of unethical, even harmful, conduct in medicine and other endeavors for which we do not pass a federal criminal law. Why is a federal law necessary to prevent human cloning but not these other actions?

In its zeal to make a strong moral statement against human cloning now, the NBAC has greatly overestimated the harms from cloning and the likelihood that they would occur. It has also underestimated the costs of a federal criminal solution. Its recommendations might be good politics, and even a commendable reminder of the interests of children. As a matter of policy, however, they are difficult to justify, and may set an undesirable precedent for dealing with future reproductive issues.

THE NBAC POSITION

In arriving at its recommendation for a federal legislative ban on human cloning, the NBAC first considered several other policy options for dealing with the safety concerns which it identified. One option, continuing the moratorium on federal funding for the creation of a child using somatic cell nuclear transfer, would not prevent "the significant risks to the child's health posed by this technology" because private sector actors would still be free to clone without federal funds.[23]

A second option—appealing to the private sector voluntarily to adhere to the President's call for a moratorium on somatic cell nuclear transfer cloning— might generate substantial compliance, but would not "deter the occasional use of somatic cell nuclear transfer cloning."[24] The Report noted that if there is a demand for a service, professionals would step forward to provide it. In addition, professional societies in the infertility field had not condemned efforts to clone a child by somatic cell nuclear transfer at the time of the Report, as had the American Medical Association.[25] Even if they had, professional bodies lack meaningful sanctions to police their members, and malpractice remedies would not provide a strong deterrent.

A third option, extending human subjects protection legislatively to private as well as publicly-funded research, would require reliance on the existing system of institutional review boards, which has been attacked in some quarters as insufficiently protective of human subjects. Also, it would not apply to therapeutic or clinical uses of cloning that are not conducted as part of a research protocol or other investigative study.[26]

The fourth policy option—a federal legislative ban—has the advantage over state bans of comprehension and clarity. If significant penalties were attached, it would increase "the deterrent effect enormously as compared to that offered by the other options," and is thus much more likely to "deter dangerous or premature efforts at cloning."[27] It would also deter commercial interests from entering this field, thus preventing the emergence of a structural force in favor of premature clinical applications.[28]

The report did recognize some drawbacks to this last option. It would prevent some states from taking a more liberal position and hinder state experimentation in how best to deal with this technology. It would also make it difficult to change policy as more information developed, because likely users of cloning may be too few to lobby effectively for change, thus leaving them without recourse to this technique even if later research shows that it is safe and effective in humans. To avoid this problem, a legislative ban should include a sunset provision to "ensure an opportunity to re-examine any judgment made today about the implications of somatic cell nuclear transfer cloning of human beings."[29] This would enable society to change course as scientific information accumulates and public discussion continues.

PROBLEMS AND CRITIQUE

Most striking about the NBAC position is the lengths to which the Commission is willing to go to prevent any use of somatic cell cloning at the present time. According to the NBAC, the harm of premature use or experimentation is so great, and the likelihood that scientists and couples would seek to clone prematurely so strong, that only a federal criminal law is sufficient to prevent it.

The idea that federal criminal law is needed to prevent potential harmful or unethical uses of a reproductive or other medical innovation is unique in bioethics. Most federal bioethical regulation has occurred through the federal funding power, not through the use of direct criminal sanctions. For example, the extensive federal regulation of human subjects research has occurred in the guise of the conditional spending power.[30] It is not a crime now to conduct human subject research without IRB review or compliance with other federal regulations, though such practices might lead to withdrawal of federal funds or independently violate state law. The Patient Self-Determination Act of 1990, which requires hospitals to inform patients of their rights to refuse medical treatment and to issue advance directives, is based on the federal power to condition Medicare spending on compliance with certain conditions.[31] Although failure to inform patients of their rights could lead to unnecessary suffering, criminal penalties were not deemed necessary to protect those interests. Nor are criminal sanctions attached to violations of the reporting requirements of the Fertility Clinic Success Rate and Laboratory Certification Act of 1992, even though false reports could cause great suffering to infertile couples who relied on them in choosing a fertility clinic.[32]

Indeed, many harmful and dangerous practices now occur in medicine without federal criminal sanctions. Physician assisted suicide and active euthanasia, which doubtlessly occur to some extent, directly harm patients but there is no federal criminal law against it. Many surgical procedures are performed on patients without adequate review of safety, yet there are no federal criminal sanctions against them. One exception is a law that makes abortion with intent to provide fetal tissue for transplant a crime.[33] This provision was added as part of a legislative compromise to enable federal funding of fetal tissue transplantation research to occur, and is arguably unconstitutional.[34]

Of course, the fact that federal criminal law has rarely been used in the bioethical area is no argument against using it now if there is a strong case for doing so. But it is precisely such a strong case that is lacking. As noted, it is highly unlikely that couples or doctors would be interested in using cloning if there were a serious risk of poor physical outcomes. The assumption that doctors and patients would be so driven by the desire to clone and to profit from doing so that they would ignore the lack of evidence showing safety and efficacy is insulting to the vast majority of doctors and fertility patients. Indeed, it views the fertility industry as willing to do anything for money, and parents wishing to choose a child's DNA as so confused or misguided about the role of genes in forming a person that they would ignore the current absence of data showing that the procedure is safe.

Still, proponents could argue that a federal ban at this time would serve the useful purpose of reassuring the public and reminding people of the great importance of the child's welfare in using genetic selection techniques. Indeed, they might also point out that those goals would be achieved at relatively little cost, for cloning has barely been done in sheep and considerable research remains before it would be done in humans. Few people would be hurt by the ban at this time because few people would legitimately be interested in cloning a child until it had been shown to be safe and effective in animal and laboratory research. In three to five years, when the ban will have lapsed or be up for renewal, cloning can be permitted if the record shows that it safely and effectively serves valid reproductive needs.

But the view that an unneeded federal law that is enacted for symbolic or political reasons is costless is itself highly dubious. If cloning quickly turns out to be a safe, effective, and desirable way for some couples to rear biologically related children, a time-limited ban will exact a price from couples who have legitimate reasons for wishing to clone. In addition, there is the danger that once Congress is moved to outlaw cloning, it may attach no time limit at all, or place the burden on proponents of cloning to have the policy changed. There is also the danger that the ban on human cloning will include a ban on human cloning embryo research. After all, if cloning itself is banned, some legislators could argue that the embryo research necessary to develop cloning also be banned. This could prevent the development not only of cloning and other genetic selection techniques, but also other kinds of medically important research.

There is also an important question of federalism here. Even if criminal sanctions were warranted, it does not follow that the federal government should enact them. We generally leave matters of family, procreation, and protection of children to the states. Only if the states are inadequate to the task is federal legislation sought. Even then, out of respect for state discretion, federal action usually occurs as an exercise of the conditional spending power, and not federal criminal law directly preempting state action.[35] Indeed, this is one of the advantages of the federal system—it allows different states to act as laboratories that experiment with different approaches, from which other states can benefit. Human cloning would appear to be a practice about which the states might have differing views, just as members of the public do. It would appear to be neither so imminent or so likely to be used in harmful ways that federal intervention from the inception is necessary.

The importance of federalism concerns is bolstered by recent Supreme Court decisions striking down federal legislation because of an insufficient commerce clause connection or intrusion on other state powers.[36] Whether or not a federal ban on cloning would survive scrutiny under these rulings, it runs counter to the increasingly common practice of leaving things with the states whenever possible.

Finally, the recommended ban sets an unwise precedent for invoking federal criminal law to settle a bioethics issue. If a federal law is justified here then it is justified for many other reproductive and genetic situations. For example, the same principles could be invoked to make it a crime for couples not to undergo genetic screening prior to conception, or to refuse prenatal testing after conception. It could also lead to a federalization of rules for assisted reproduction, or for future developments in gene therapy and genetic alteration. If so, the NBAC's cloning report will truly have been historical. It would have started the process by which federal criminal law is used to regulate the use of all new reproductive and genetic selection technologies.

In sum, the NBAC has assumed a worst case scenario and recommended the strongest federal policy response—the enactment of a criminal law to ban all such uses. It has not shown that the risks are so great nor so likely to occur as to justify criminal law at the federal level, nor is it sufficiently sensitive to the procreative liberty and federalism costs of such an approach. In its desire to condemn human cloning now, it has recommended a policy that needs much stronger support than the NBAC Report gives.

CLONING POLICY IN THE FUTURE

The NBAC's main recommendation was directed to short-run policy for human cloning. Its call for a time-limited federal ban might be viewed as holding the fort until we have had more time to sort out the issues and determine a more considered policy.

A key question in any later evaluation concerns who has the burden of establishing that cloning or other forms of genetic selection are a net good or a net harm. The NBAC assumed that the proponents now have the burden, because nuclear transfer cloning is new and unsafe. However, if animal and laboratory studies show that human cloning is safe and effective, the burden of proof should remain on those who seek to use it, as some might argue if a federal ban is still in effect.

Our ethical, legal, and social commitment to reproductive and family liberty should place the burden on opponents to show that family-centered uses of cloning are not truly procreative, or that they impose such a high risk of severe harm that they should be prohibited.[37] Such an approach is necessary to give procreative liberty and family autonomy its due. Speculation, hypothetical harms, and moral objection alone are not a sufficient basis for limiting coital reproduction. They should not be sufficient to limit the use of noncoital and genetic selection techniques essential to a couple seeking healthy, biologically-related children to rear.

This means that future inquiries into the ethics and public policy issues raised by human cloning and other genetic selection techniques should begin with an examination of their role in helping families have healthy children to rear. As I argue elsewhere, a plausible case exists for viewing cloning in some circumstances as an essential way for a married couple intent on gestating and rearing children to accomplish that goal.[38] Once the connection with procreative choice is established, proper respect for family and reproductive liberty requires that opponents of the practice establish that legitimate family uses would produce such great harm to others that prohibition or regulation is justified.

In assessing harms from cloning and other genetic selection techniques, a much more careful analysis than occurred in Chapter 4 of the NBAC report will be essential. It will not be sufficient to assert that abuses *could* occur, or that a child's individuality or autonomy *could* be affected, or that problems about family lineage *could* result.[39] Nor does it suffice to note that some people will always have moral objections to cloning or to other techniques. The important question is whether these concerns constitute such serious or severe harms and are so likely to occur in most uses of the technique, that all uses can be banned, even clearly ethical and valid uses.

A careful analysis of the likely uses of cloning, the harms that are likely, and the ways that those harms might be minimized will provide a much better grounding for public policy than assessments guided by moral objection alone or fears of possible harm that are not likely in most cases. These are the questions to be addressed if we are to come to satisfactory terms with the use of cloning and other forms of genetic selection to form families.

---

1. A day later, he called for a moratorium on private sector cloning as well.
2. *Cloning Human Beings: Report and Recommendations of the National Bioethics Advisory Commission,* Rockville, Maryland, 1997. (Hereinafter "Report")
3. Bills to outlaw human cloning had also been introduced in eight states. See Report, p. 104.
4. Id. at 102–103. The joint communique of the eight nation economic summit meeting in Denver in June, 1997 included an agreement by each country to prohibit human cloning for the creation of children. David S. Cloud, "Achievements at Summit of 8 Often Less Than Meets the Eye," *Chicago Tribune,* June 23, 1997 at 4.
5. To give but one example, one sentence appears to state the outrageous statement that one is immature until one has reproduced: "Without reproduction one remains a child and perhaps a sibling. With reproduction—or . . . adoption—one becomes a parent, taking on responsibilities for another that necessarily require abandoning some of the personal freedoms enjoyed before." Report at 77. If another meaning was intended, as is likely, the sentence should have been clarified.
6. The introduction usefully lays out the sudden emergence of controversy over cloning. The scientific chapter is important background for anyone wanting to know basic scientific facts and potential uses of cloning. The religious chapter surveys major faiths to see if they have a position on cloning, and finds that while some are against it, others are not. The legal and policy chapter discusses whether existing state or federal law deals with cloning or cloning research, addresses possible constitutional barriers to regulating cloning, and identifies several options for a federal policy response.
7. What exactly such a law would prohibit will depend on the text adopted. If the law's intent is to penalize only actions intended to bring about the birth of a child, much human embryo cloning research might still be permitted. Strictly speaking, transferring embryos to the uterus to see how effectively they implant, and if they do, to abort them at 12 weeks, would also not violate such a statute. If a subject then changed her mind and refused to abort, the statute still would not have been technically violated because the intent at time of transfer was not to bring about the birth of a cloned child.
8. "The prospect of creating children through somatic cell nuclear transfer has elicited widespread concern because of fears about harms to the children who are born as a result. There are concerns about physical harms from manipulation of ova, nuclei,

and embryos, which are parts of the technology, and about possible psychological harms, such as a diminished sense of individuality and personal autonomy. . . . *Virtually all people agree that the current risks of physical harm to children associated with somatic cell nuclear transplantation cloning justify a prohibition at this time on such experimentation."* (emphasis supplied) Report, p. 63.

9. Because the Report does not always make clear whether it is talking about experimental or about clinical uses, the reader is reminded that procedures that are "experimental" in the sense that data about their efficacy is not known may still be used with a clinical or therapeutic intent and thus not constitute experimentation. That is, the primary purpose may be to help the couple to have a child rather than to obtain knowledge.

10. Report, p. 64.

11. At this point the Report cites the Nuremberg Code on human experimentation as also supporting the notion that avoidance of physical or psychological harm is the standard for assessing the ethics of research. The Report overlooks the fact that an unproven therapy might be used with clinical intent, and thus not be intended as research. See note 9 supra. It also is an inaccurate statement of Nuremberg principles. Harm or the risk of harm usually occurs in human subjects research. If the subject fully and freely consented to that risk, harm-producing research is not then unethical.

12. Report at p. 65.

13. Id.

14. Id.

15. The inclusion of fetuses might be a bow in the direction of anti-abortion interests. As a separate ethical statement or basis for prohibition, the claim is highly dubious. If fetuses have no rights prior to viability, then aborting them cannot be unethical except on symbolic grounds. See John A. Robertson, "Symbolic Issues In Embryo Research," Hastings Center Report, Jan./Feb. 1995, 37–38. A couple willing to risk a high rate of miscarriage in order to have a child with particular DNA is not necessarily acting unethically, much less committing an act that warrants federal criminal sanctions.

16. See, e.g., *Johnson v. State,* 602 So.2d 1288 (Fla. 1992).

17. While the courts have allowed parents to recover damages on their own behalf for wrongful birth, they have invariably rejected wrongful life suits by handicapped children. *Smith v. Cote,* 513 A.2d 342 (N.H. 1986); *Becker v. Shwartz,* 386 N.E.2d 807 (N.Y. 1978); *Nelson v. Krusen,* 678 S.W.2d 918 (Tx. 1984). However, the California, Washington, and New Jersey Supreme Courts have allowed children to recover special but not general damages on a claim of wrongful life in situations in which their parents were able to recover both special and general damages for the child's birth. *See Turpin v. Sortini,* 31 Cal.3d 220, 643 P.2d 954, 182 Cal. Rptr. 337 (1982); *Procanik v. Cillo,* 97 N.J. 339, 478 A.2d 755 (1984); *Harverson v. Parke-*

Davis, 98 Wash.2d 460, 656 P.2d 483 (1983). Perhaps the results in these cases are defensible as a means to assure that the tortfeasor internalizes the full costs of the tort. However, the cases are confused about the meaning of wrongful life and inconsistent within their own terms, because they do not allow the child recovery for general damages. If the child has been wronged by being born, then such damages also should be awarded.

18. It concludes the paragraph with the statement that "Even the originator of this line of analysis rejects this conclusions (sic)." Report at p. 66. A footnote to this sentence cites Derek Parfit and the nonidentity problem, overlooking the fact that Parfit's claim depends on the assumption that the couple could easily have a healthy child instead. Yet the use of cloning and other controversial techniques is likely to be sought precisely because the couple cannot easily have a healthy child. Because the numbers cannot be kept constant, Parfit has therefore not shown that the argument against the child being harmed from existence alone is invalid.

19. It thus makes no claim about a right to be born when one's life will be a net benefit to that person, for there is no one yet born to have rights.

20. If the child were already in existence, one would be ethically permitted or even obligated to prevent further suffering by either passive or active euthanasia (the legal permissibility of such measures is another matter).

21. Indeed, in concluding that nonexistence is better than existence as a clone and that cloning should therefore not occur, the report makes the very metaphysical error it had accused the other side of making. For the report compares existence with nonexistence before any child is born, whereas the position it criticizes compares the two states only after the child is born.

22. In my view a more fruitful way to approach the issue is through asking whether the intentions of the parents regarding gestation and rearing falls within societal understandings and conventions about procreation and family. If so, they should be permitted knowingly to risk having children through cloning and other techniques that could lead to undesirable physical or psychological conditions.

23. Report, p. 96.

24. Id. at 98.

25. The NBAC apparently overlooked the opposition of the American Society of Reproductive Medicine to cloning at this time, a position that was presented to it during its public hearing on March 14, 1997. See also, "ASRM Statement on Human Cloning Through Nuclear Transplantation," Press Release, American Society of Reproductive Medicine, Birmingham, Alabama, March 5, 1997.

26. Report at 99. Again the NBAC is imprecise in its language. In talking about "experimental use of this technology that is offered in a therapeutic or other non-research guise" it must mean "experimental" in the sense of unproven or unestablished as safe and effective. For if the therapeutic intent is experimental, at the same time

that it is being used for clinical therapy, then it would constitute research and be subject to IRB review. See notes 9, 11 supra.

27. Report, p. 101.

28. The report also noted that a ban would quell anxieties about the use of purely molecular and cellular techniques, which offer great promise for medical and scientific advance without posing the same ethical issues that arise in creating a child.

29. Report, p. 101.

30. John A. Robertson, "The Law of Institutional Review Boards," 26 U.C.L.A. Law Rev. 484 (1979).

31. Omnibus Budget Reconciliation Act of 1990, Pub.L. No. 101–508 # 4206, 4751, 104 Stat. 1388, 1388–115 to 117, 1388–204 to 206. Codified in scattered sections of 42 U.S.C.

32. Public Law 12–493 (H.R. 4773), 24 October 1992.

33. National Institutes of Health Revitalization Act of 1993, Pub.L. No. 103–43, #112, 107 Stat. 131 (codified at 42 U.S.C. #289g-2 (Supp. 1994)).

34. John A. Robertson, "Abortion to Obtain Fetal Tissue for Transplant," 27 Suffolk U. L. Rev. 1359 (1993).

35. A typical example is "Megan's law," which requires states to pass laws that give communities notice of the release of sex offenders in order to receive certain kinds of federal funding. Jacob Wetterling Crimes Against Children and Sexually Violent Offender Registration Program #170101, 42 U.S.C. #14071 (1994). Federal rules for withholding treatment from handicapped newborns are imposed as a condition on the receipt of federal funds for child abuse programs. Child Abuse Amendments of 1984 (Pub.L. 98–457) #307, 42 U.S.C. #10406 (1994). On the other hand, the federal battered woman law was a direct criminal imposition. Violence Against Women Act of 1994 #40302, 42 U.S.C. #13981 (1994). But see *Brzonkala v. Virginia Polytechnic and State University*, 935 F.Supp. 779 (1996) (Violence Against Women Act not within authority of Congress under the commerce clause or under the enforcement clause of the Fourteenth Amendment).

36. See *United States v. Lopez*, 514 U.S. 549 (1995); *Printz v. United States*, 65 U.S. Law Week 4731.

37. For arguments supporting this claim, see John A. Robertson, "Children of Choice: Freedom and New Reproductive Technologies" (1994).

38. John A. Robertson, "Liberty, Identity, and Human Cloning" (forthcoming).

39. Too often the NBAC report falls into overbroad or simplistic slippery slope reasoning, i.e., because some people think that cloning is wrong or that it will be misused, all cloning should be banned. This is not an adequate basis for infringing other fundamental freedoms, as is evident in the First Amendment context. It should not suffice to limit assisted reproductive and genetic selection techniques that enable married couples and individuals to have and rear biologically-related children.

DOLLY'S FASHION

AND LOUIS'S PASSION

STEPHEN JAY GOULD

Stephen Jay Gould, one of America's most famous scientists, is a professor of paleontology and zoology at Harvard and New York Universities, 1998 president of the American Academy for the Advancement of Science, and author of fifteen books for general audiences. He writes on topics such as the panda's thumb and Tyrannosaurus dinosaurs. Since 1974, he has written a monthly column for the magazine Natural History *without missing an issue.*

Gould's most famous scientific work describes, based on fossils, the evolution of Gryphaea, *a kind of land snail. He is also famous for his attacks on genetic reductionism.*

Gould has always argued that, although biblical inerrantists are incorrect to argue that evolution is false, evolution is not as simple as Darwin thought. Gould believes that evolution occurs by fits and starts in a "punctuated equilibrium."

In the piece that follows, Gould argues that facts about conjoined twins are the best predictors of who two humans would be who were originated at different times from the same genes. He believes that such time-separated, genetically identical twins would not be very similar in talent, personality, or achievement. At the end of the essay, he discusses the work of his friend Frank Sulloway on birth order of children in families as a predictor of later achievement. Although he disagrees with some of Sulloway's conclusions, Gould believes that Sulloway's work shows the profound effect of the environment on human personality and achievement.

NOTHING CAN BE MORE FLEETING OR CAPRICIOUS THAN FASHION. What, then, can a scientist, committed to objective description and analysis, do with such a haphazardly moving target? In a classic approach, analogous to standard advice for preventing the spread of an evil agent ("kill it before it multiplies"), a scientist might say, "quantify before it disappears."

Francis Galton, Charles Darwin's charmingly eccentric and brilliant cousin, and a founder of the science of statistics, surely took this prescription to heart. He once decided to measure the geographic patterning of female beauty. He attached a piece of paper to a small wooden cross that he could carry, unobserved, in his pocket. He held the cross at one end in the palm of his hand and, with a needle secured between thumb and forefinger, made pinpricks on the three remaining projections (the two ends of the crossbar and the top).

He would rank every young woman he passed on the street into one of three categories—as beautiful, average, or substandard (by his admittedly subjective preferences)—and he would then place a pinprick for each woman into the designated domain of his cross. After a hard day's work, he tabulated the relative percentages by counting pinpricks. He concluded, to the dismay of Scotland, that beauty followed a simple trend from north to south, with the highest proportion of uglies in Aberdeen and the greatest frequency of lovelies in London.

Some fashions (tongue piercings, perhaps?) flower once and then disappear, hopefully forever. Others swing in and out of style, as if fastened to the end of a pendulum. Two foibles of human life strongly promote this oscillatory mode. First, our need to create order in a complex world begets our worst mental habit: dichotomy, or our tendency to reduce an intricate set of subtle shadings to a choice between two diametrically opposed alternatives (each with moral weight and therefore ripe for bombast and pontification, if not outright warfare): religion versus science, liberal versus conservative, plain versus fancy, *Roll Over Beethoven* versus the *Moonlight Sonata*. Second, many deep questions about our livelihoods, and the fates of nations, truly have no answers—so we cycle the presumed alternatives of our dichotomies, one after the other, always hoping that, this time, we will find the nonexistent key.

Among oscillating fashions governed primarily by the swing of our social pendulum, no issue could be more prominent for an evolutionary biologist, or more central to a broad range of political questions, than genetic versus environmental sources of human abilities and behaviors. This issue has been falsely dichotomized for so many centuries that English even features a mellifluous linguistic contrast for the supposed alternatives: nature versus nurture.

As any thoughtful person understands, the framing of this question as an either-or dichotomy verges on the nonsensical. Both inheritance and upbringing matter in crucial ways. Moreover, an adult human being, built by interaction of these (and other) factors, cannot be disaggregated into separate components with attached percentages. It behooves us all to grasp why such

common claims as "intelligence is 30 percent genetic and 70 percent environmental" have no sensible meaning at all and represent the same kind of error
as the contention that all overt properties of water may be revealed by noting
an underlying construction from two parts of one gas mixed with one part of
another.

Nonetheless, a preference for either nature or nurture swings back and
forth into fashion as political winds blow and as scientific breakthroughs grant
transient prominence to one or another feature in a spectrum of vital influences. For example, a combination of political and scientific factors favored
an emphasis upon environment in the years just following World War II: an
understanding that Hitlerian horrors had been rationalized by claptrap genetic
theories about inferior races; the domination of psychology by behaviorist theories. Today, genetic explanations are all the rage, fostered by a similar mixture
of social and scientific influences: for example, the rightward shift of the political pendulum (and the cynical availability of "you can't change them, they're
made that way" as a bogus argument for reducing expenditures on social programs) and an overextension to all behavioral variation of genuinely exciting
results in identifying the genetic basis of specific diseases, both physical and
mental.

Unfortunately, in the heat of immediate enthusiasm we often mistake transient fashion for permanent enlightenment. Thus, many people assume that
the current popularity of genetic explanation represents a final truth wrested
from the clutches of benighted environmental determinists of previous generations. But the lessons of history suggest that the worm will soon turn again.
Since both nature and nurture can teach us so much—and since the fullness
of our behavior and mentality represents such a complex and unbreakable combination of these and other factors—a current emphasis on nature will no
doubt yield to a future fascination with nurture as we move toward better
understanding by lurching upward from one side to another in our quest to
fulfill the Socratic injunction: know thyself.

In my Galtonian desire to measure the extent of current fascination with
genetic explanations (before the pendulum swings once again and my opportunity evaporates), I hasten to invoke two highly newsworthy items of recent
months. The subjects may seem quite unrelated—Dolly, the cloned sheep, and
Frank Sulloway's book on the effects of birth order upon human behavior—but
both stories share a common feature offering striking insight into the current
extent of genetic preferences. In short, both stories have been reported almost
entirely in genetic terms, but both cry out (at least to me) for a reading as proof
of strong environmental influences. Yet no one seems to be drawing (or even
mentioning) this glaringly obvious inference. I cannot imagine that anything
beyond current fashion for genetic arguments can explain this puzzling silence.

I am convinced that exactly the same information, if presented twenty years ago in a climate favoring explanations based on nurture, would have been read primarily in this opposite light. Our world, beset by ignorance and human nastiness, contains quite enough background darkness. Should we not let both beacons shine all the time?

Dolly must be the most famous sheep since John the Baptist designated Jesus in metaphor as "Lamb of God, which taketh away the sin of the world" (John: 1:29). She has certainly edged past the pope, the president, Madonna, and Michael Jordan as the best-known mammal of the moment. And all this for a carbon copy, a Xerox! I don't intend to drip cold water on this little lamb, cloned from a mammary cell of her mother, but I remain unsure that she's worth all the fuss and fear generated by her unconventional birth.

When one reads the technical article describing Dolly's manufacture ("Viable Offspring Derived from Fetal and Adult Mammalian Cells," by I. Wilmut, A. E. Schnieke, J. McWhir, A. J. Kind, and K. H. S. Campbell, *Nature*, February 27, 1997), rather than the fumings and hyperbole of so much public commentary, one can't help feeling a bit underwhelmed and left wondering whether Dolly's story tells less than meets the eye.

I don't mean to discount or underplay the ethical issues raised by Dolly's birth (and I shall return to this subject in a moment), but we are not about to face an army of Hitlers or even a Kentucky Derby run entirely by genetically identical contestants (a true test for the skills of jockeys and trainers). First, Dolly breaks no theoretical ground in biology, for we have known how to clone in principle for at least two decades, but had developed no techniques for reviving the full genetic potential of differentiated adult cells. (Still, I admit that a technological solution can pack as much practical and ethical punch as a theoretical breakthrough. I suppose one could argue that the first atomic bomb only realized a known possibility.)

Second, my colleagues have been able to clone animals from embryonic cell-lines for several years, so Dolly is not the first mammalian clone, but only the first clone from an adult cell. Wilmut and colleagues also cloned sheep from cells of a nine-day embryo and a twenty-six-day fetus—and had much greater success. They achieved fifteen pregnancies (although not all proceeded to term) in thirty-two recipients (that is, surrogate mothers for transported cells) of the embryonic cell-line, five pregnancies in sixteen recipients of the fetal cell-line, but only Dolly (one pregnancy in thirteen tries) for the adult cell-line. This experiment cries out for confirming repetition. (Still, I allow that current difficulties will surely be overcome, and cloning from adult cells, if doable at all, will no doubt be achieved more routinely as techniques and familiarity improve.)

Third, and more seriously, I remain unconvinced that we should regard

Dolly's starting cell as adult in the usual sense of the term. Dolly grew from a cell taken from the "mammary gland of a six-year-old ewe in the last trimester of pregnancy" (to quote the technical article of Wilmut, et al.). Since the breasts of pregnant mammals enlarge substantially in late stages of pregnancy, some mammary cells, although technically adult, may remain unusually labile or even "embryolike" and thus able to proliferate rapidly to produce new breast tissue at an appropriate stage of pregnancy. Consequently, we may be able to clone only from unusual adult cells with effectively embryonic potential, and not from any stray cheek cell, hair follicle, or drop of blood that happens to fall into the clutches of a mad Xeroxer. Wilmut and colleagues admit this possibility in a sentence written with all the obtuseness of conventional scientific prose, and therefore almost universally missed by journalists: "We cannot exclude the possibility that there is a small proportion of relatively undifferentiated stem cells able to support regeneration of the mammary gland during pregnancy."

But if I remain relatively unimpressed by achievements thus far, I do not discount the monumental ethical issues raised by the possibility of cloning from adult cells. Yes, we have cloned fruit trees for decades by the ordinary process of grafting—and without raising any moral alarms. Yes, we may not face the evolutionary dangers of genetic uniformity in crop plants and livestock, for I trust that plant and animal breeders will not be stupid enough to eliminate all but one genotype from a species and will always maintain (as plant breeders do now) an active pool of genetic diversity in reserve. (But then, I suppose we should never underestimate the potential extent of human stupidity—and agricultural seed banks could be destroyed by local catastrophes, while genetic diversity spread throughout a species guarantees maximal evolutionary robustness.)

Nonetheless, while I regard many widely expressed fears as exaggerated, I do worry deeply about potential abuses of human cloning, and I do urge a most open and thorough debate on these issues. Each of us can devise a personal worst-case scenario. Somehow, I do not focus upon the specter of a future Hitler making an army of ten million identical robotic killers, for if our society ever reaches a state in which such an outcome might be realized, we are probably already lost. My thoughts run to localized moral quagmires that we might actually have to face in the next few years (for example, the biotech equivalent of ambulance-chasing slimeballs among lawyers—a hustling little firm that scans the obits for reports of dead children and then goes to grieving parents with the following offer: "So sorry for your loss, but did you save a hair sample? We can make you another for a mere fifty thou").

However, and still on the subject of ethical conundrums, but now moving to my main point about current underplaying of environmental sources for human behaviors, I do think that the most potent scenarios of fear, and the most fretful ethical discussions on late-night television, have focused on a non-existent problem that all human societies solved millennia ago. We ask: Is a clone an individual? Would a clone have a soul? Would a clone made from my cell negate my unique personhood?

May I suggest that these endless questions—all variations on the theme that clones threaten our traditional concept of individuality—have already been answered empirically, even though public discussion of Dolly seems blithely oblivious to this evident fact. We have known human clones from the dawn of our consciousness. We call them identical twins—and they are far better clones than Dolly and her mother. Dolly shares only nuclear DNA with her mother's mammary cell, for the nucleus of this cell was inserted into an embryonic stem cell (whose own nucleus had been removed) of a surrogate female. Dolly then grew in the womb of this surrogate.

Identical twins share at least four additional (and important) properties that differ between Dolly and her mother. First, identical twins also house the same mitochondrial genes. (Mitochondria, the "energy factories" of cells, contain a small number of genes. We get our mitochondria from the cytoplasm of the egg cell that made us, not from the nucleus formed by the union of sperm and egg. Dolly received her nucleus from her mother, but her egg cytoplasm, and hence her mitochondria, from her surrogate.) Second, identical twins share the same set of maternal gene products in the egg. Genes don't grow embryos all by themselves. Egg cells contain protein products of maternal genes that play a major role in directing the early development of the embryo. Dolly has her mother's nuclear genes, but her surrogate's gene products in the cytoplasm of her founding cell.

Third—and now we come to explicitly environmental factors—identical twins share the same womb. Dolly and her mother gestated in different places. Fourth, identical twins share the same time and culture (even if they fall into the rare category, so cherished by researchers, of siblings separated at birth and raised, unbeknownst to each other, in distant families of different social classes). The clone of an adult cell matures in a different world. Does anyone seriously believe that a clone of Beethoven would sit down one day to write a Tenth Symphony in the style of his early-nineteenth-century forebear?

So identical twins are truly eerie clones—ever so much more alike on all counts than Dolly and her mother. We do know that identical twins share massive similarities not only of appearance but also in broad propensities and detailed quirks of personality. Nonetheless, have we ever doubted the personhood of each member in a pair of identical twins? Of course not. We know that identical twins are distinct individuals, albeit with peculiar and extensive similarities. We give them different names. They encounter divergent experiences and fates. Their lives wander along disparate paths of the world's complex vagaries. They grow up as distinctive and undoubted individuals, yet they stand forth as far better clones than Dolly and her mother.

Why have we overlooked this central principle in our fears about Dolly? Identical twins provide sturdy proof that inevitable differences of nurture guarantee the individuality and personhood of each human clone. And since any future human Dolly must differ far more from her progenitor (in both the nature of mitochondria and maternal gene products and the nurture of different wombs and surrounding cultures) than any identical twin diverges from her sibling clone, why ask if Dolly has a soul or an independent life when we have never doubted the personhood or individuality of much more similar identical twins?

Literature has always recognized this principle. The Nazi loyalists who cloned Hitler in *The Boys from Brazil* also understood that they had to maximize similarities of nurture as well. So they fostered their little Hitler babies in families maximally like Adolf's own dysfunctional clan—and not one of them grew up anything like history's quintessential monster. Life, too, has always verified this principle. Eng and Chang, the original Siamese twins and the closest clones of all, developed distinct and divergent personalities. One became a morose alcoholic, the other remained a benign and cheerful man. We may not think much of the individuality of sheep in general (for they do set our icon of blind following and identical form as they jump over fences in mental schemes of insomniacs), but Dolly will grow up to be as unique and as ornery as any sheep can be.

A recent book by my friend Frank Sulloway also focuses on themes of nature and nurture. He fretted over, massaged, and lovingly shepherded it toward publication for more than two decades. *Born to Rebel* documents a crucial effect of birth order in shaping human personalities and styles of thinking. Firstborns, as sole recipients of parental attention until the arrival of later children, and as more powerful (by virtue of age and size) than their subsequent siblings, tend to cast their lot with parental authority and with the advantages of incumbent strength. They tend to grow up competent and confident, but also conservative and unlikely to favor quirkiness or innovation. Why threaten an existing structure that has always offered you clear advantages over siblings? Later children, however, are (as Sulloway's title proclaims) born to rebel. They must compete against odds for parental attention long focused primarily elsewhere. They must scrap and struggle and learn to make do for themselves. Laterborns therefore tend to be flexible, innovative, and open to change. The business and political leaders of stable nations may be overwhelmingly firstborns, but the revolutionaries who have discombobulated our cultures and restructured our scientific knowledge tend to be laterborns. Frank and I have been discussing his thesis ever since he began his studies. I thought (and suggested) that he should have published his results twenty years ago. I still hold this opinion, for while I greatly admire his book and do recognize that such a long gestation allowed Frank to strengthen his case by gathering and refining his data, I also believe that he became too committed to his central thesis and tried to extend his explanatory umbrella over too wide a range, with arguments that sometimes smack of special pleading and tortured logic.

Sulloway defends his thesis with statistical data on the relationship of birth order and professional achievement in modern societies—and by interpreting historical patterns as strongly influenced by characteristic differences in behavior of firstborns and laterborns. I found some of his historical arguments fascinating and persuasive when applied to large samples but often uncomfortably overinterpreted in attempts to explain the intricate details of individual lives (for example, the effect of birth order on the differential success of Henry VIII's various wives in overcoming his capricious cruelties).

In a fascinating case, Sulloway chronicles a consistent shift in relative percentages of firstborns among successive groups in power during the French Revolution. The moderates initially in charge tended to be firstborns. As the revolution became more radical, but still idealistic and open to innovation and free discussion, laterborns strongly predominated. But when control then passed to the uncompromising hardliners who promulgated the Reign of Terror, firstborns again ruled the roost. In a brilliant stroke, Sulloway tabulates the birth orders for several hundred delegates who decided the fate of Louis XVI in the National Convention. Among hardliners who voted for the guillotine, 73 percent were firstborns; but of those who opted for the compromise of conviction with pardon, 62 percent were laterborns. Since Louis lost his head by a margin of one vote, an ever so slightly different mix of birth orders among delegates might have altered the course of history.

Since Frank is a good friend and since I have been at least a minor midwife to this project over two decades (although I don't accept all details of his thesis), I took an unusually strong interest in the delayed birth of *Born to Rebel.* I read the text and all the prominent reviews that appeared in many newspapers and journals. And I have been puzzled—stunned would not be too strong a word—by the total absence from all commentary of the simplest and most evident inference from Frank's data, the one glaringly obvious point that everyone should have stressed, given the long history of issues raised by such information.

Sulloway focuses nearly all his interpretation on an extended analogy (broadly valid in my judgment, but overextended as an exclusive device) between birth order in families and ecological status in a world of Darwinian competition. Children vie for limited parental resources, just as individuals struggle for existence (and ultimately for reproductive success) in nature. Birth orders place children in different "niches," requiring disparate modes of competition for maximal success. While firstborns shore up incumbent advantages, laterborns must grope and grub by all clever means at their disposal—leading to the divergent personalities of stalwart and rebel. Alan Wolfe, in my favorite negative review of Sulloway's book from the *New Republic* (December 23, 1996) writes: "Since firstborns already occupy their own niches, laterborns, if they are to be noticed, have to find unoccupied niches. If they do so successfully, they will be rewarded with parental investment." (Jared Diamond stresses the same theme in my favorite positive review from the *New York Review of Books*, November 14, 1996.)

As I said, I am willing to go with this program up to a point. But I must also note that the restriction of commentary to this Darwinian metaphor has diverted attention from the foremost conclusion revealed by a large effect of birth order upon human behavior. The Darwinian metaphor smacks of biology; we also erroneously think of biological explanations as intrinsically genetic (an analysis of this common fallacy could fill an essay or an entire book). I suppose that this chain of argument leads us to stress whatever we think that Sulloway's thesis might be teaching us about "nature" (our preference, in any case, during this age of transient fashion for genetic causes) under our erroneous tendency to treat the explanation of human behavior as a debate between nature and nurture.

But consider the meaning of birth-order effects for environmental influences, however unfashionable at the moment. Siblings differ genetically of course, but no aspect of this genetic variation correlates in any systematic way with birth order. Firstborns and laterborns receive the same genetic shake within a family. Systematic differences in behavior between firstborns and laterborns cannot be ascribed to genetics. (Other biological effects may correlate with birth order—if, for example, the environment of the womb changes systematically with numbers of pregnancies—but such putative influences have no basis in genetic differences among siblings.) Sulloway's substantial birth-order effects therefore provide our best and ultimate documentation of nurture's power. If birth order looms so large in setting the paths of history and the allocation of people to professions, then nurture cannot be denied a powerfully formative role in our intellectual and behavioral variation. To be sure, we often fail to see what stares us in the face, but how can the winds of fashion blow away such an obvious point, one so relevant to our deepest and most persistent questions about ourselves?

In this case, I am especially struck by the irony of fashion's veil. As noted before, I urged Sulloway to publish this data twenty years ago, when (in my judgment) he could have presented an even better case because he had already documented the strong and general influence of birth order upon personality, but had not yet ventured upon the slippery path of trying to explain too many details with forced arguments that sometimes lapse into self-parody. If Sulloway had published in the mid-1970s, when nurture rode the pendulum of fashion in a politically more liberal age (probably dominated by laterborns!), I am confident that this obvious point about birth-order effects as proof of nurture's power would have won primary attention, rather than consignment to a limbo of invisibility.

Hardly anything in intellectual life can be more salutatory than the separation of fashion from fact. Always suspect fashion (especially when the moment's custom matches your personal predilection); always cherish fact (while remembering that an apparent "fact" may only record a transient fashion). I have discussed two subjects that couldn't be "hotter," but cannot be adequately understood because a veil of genetic fashion now conceals the richness of full explanation by relegating a preeminent environmental theme to invisibility. Thus, we worry whether the first cloned sheep represents a genuine individual at all, while we forget that we have never doubted the distinct personhood guaranteed by differences in nurture to clones far more similar by nature than Dolly and her mother—identical twins. And we try to explain the strong effects of birth order only by invoking a Darwinian analogy between family place and ecological niche, while forgetting that these systematic effects cannot have a genetic basis and therefore prove the predictable power of nurture.

So, sorry, Louis. You lost your head to the power of family environments upon head children. And hello, Dolly. May we forever restrict your mode of manufacture, at least for humans. But may genetic custom never stale the infinite variety guaranteed by a lifetime of nurture in the intricate complexity of nature—this vale of tears, joy, and endless wonder.

CLONE MAMMALS . . .

CLONE MAN?

AXEL KAHN

*Axel Kahn, director, INSERM Laboratory of Research on Genetics and Molecu-
lar Pathology, Cochin Institute of Molecular Genetics, Paris, is a distinguished
scientist known for his work on genetics, cancer, and mental retardation. He
has cloned a gene for a disorder of the central nervous system. He advises the
French National Institute for Health and Medical Research on issues in research ethics.*

THE EXPERIMENTS OF I. WILMUT ET AL. (*Nature* 385, 810; 1997)
demonstrate that mammalian oocytes can reprogramme the nuclei of somatic
cells, enabling them to become totipotent. For the moment, however, the precise conditions under which this can occur remain to be elucidated: the proportion of normal embryos obtained is very low, and we have no information on
the longevity or fecundity of the lambs produced. Therefore, the factors determining the success or failure of the technique, and the long-term development
of animals generated in this way, still need to be established.

But this is beside the main point, which is that Wilmut et al. have shown
that it is now possible to envisage cloning of adult mammals in a completely
asexual fashion. The oocyte's only involvement is the role of its cytoplasm in
reprogramming the introduced nucleus and in contributing intracellular organelles—mainly mitochondria—to the future organism.

This work will undoubtedly open up new perspectives in research in biology
and development, prompting rapid progress for example, in understanding of
the functional plasticity of the genome and chromatin during development,
and the mechanisms guaranteeing the stability of differentiated states. Another
immediate scientific question is whether a species barrier exists. Could an em-

bryo be produced, for example, by implanting the nucleus of a lamb in an enucleated mouse oocyte? Any lambs born in this way would possess a mouse mitochondrial genome.

In passing, the technique suggests that a woman suffering from a serious mitochondrial disease might in future be able to produce children free of the disease by having the nucleus of her embryo implanted in a donor oocyte (note that this process is not the same as "cloning").

But the primordial medical, philosophical and ethical question that is bound to be raised by scientists, philosophers, ethicists and the public following publication of the paper by Wilmut et al. is of the possibility, legitimacy and medical implications of human cloning using such techniques. There is no a priori reason to suspect that humans should behave very differently from other mammals where cloning is possible, and the cloning of an adult human could therefore become feasible using the techniques reported by Wilmut et al. What medical and scientific "justifications" might there be for cloning? Previous debates on human cloning have identified the preparation of immunocompatible differentiated cell lines—such as hematopoetic, neuronal and pancreatic cells—for transplantation, as one potential indication. We could imagine everyone creating a reserve of therapeutic cells that would increase their chance of being cured of various diseases, such as cancer, degenerative disorders, and viral or inflammatory diseases. Such cells could even be corrected ex vivo in the case of genetic diseases.

But the debate has in the past perhaps paid insufficient attention to the current strong social and psychological trend towards a fanatical desire for individuals not simply to have children but to ensure that these children also carry their genes, even when faced with the obstacle of sterility (or death). Achieving such biological descendance was impossible for sterile men until the development of ICSI (intracytoplasmic sperm injection). This technique allows a single sperm to be injected directly into the oocyte, whereas the only option previously available for couples where the man was sterile due to low sperm quality or levels was insemination with donor sperm.

But human descendance is not only biological, as it is in all other species, but is also emotional and cultural. The latter is of such importance that methods of inheritance where both parents' genes are not transmitted—such as adoption and insemination with donor sperm—are widely accepted without any major ethical questions being raised.

But today's society is characterized by an increasing demand for biological inheritance, as if this were the only form of inheritance worthy of the name. One reason is that, regrettably, a person's personality is increasingly perceived as being largely determined by his or her genes. Moreover, in a world where culture is increasingly internationalized and homogenized, our fellow citizens may ask themselves whether they have anything else to transmit to their children apart from their genes.

This pressure probably accounts for the wide social acceptation of ICSI, a technique which, it should be remembered, was widely made available to people at a time when experimental evidence as to its safety was still flimsy. ICSI means that men with abnormal sperm can now procreate. Moreover, a recent extension of the technique using immature spermatids means that even men who are unable to produce mature spermatozoa can now also procreate.

Going further upstream, researchers have now succeeded in fertilizing a mouse oocyte using a diploid nucleus of a spermatogonium: they find that apparently normal embryonic development occurs, at least in the early stages. But there are also severe forms of sterility—such as dysplasia or severe testicular atrophy—or indeed lesbian couples, where no male germ line exists. Will such couples also demand the right to a biological descendance?

Applying the technique used by Wilmut et al. in sheep directly to humans would yield a clone "of the father" and not a shared descendant of both the father and mother. Nevertheless, for a woman the act of carrying a fetus can be as important as being its biological mother. The extraordinary power of such "maternal reappropriation" of the embryo can be seen from the strong demand for pregnancies in post-menopausal women, and for embryo and oocyte donations to circumvent female sterility. Moreover, if cloning techniques were ever to be used, the mother would be contributing something—her mitochondrial genome. This suggests that we probably cannot exclude the possibility that the current direction of public opinion will tend to legitimize the resort to cloning techniques in cases, where, for example, the male partner in a couple is unable to produce any gametes.

The creation of human clones with the sole aim of preparing spare cell lines would from a philosophical point of view be in obvious contradiction with an ethical principle expressed by Immanuel Kant: that of human dignity. This principle demands that an individual—and I would extend this to read human life—should never be thought of only as a means, but always also as an end. The objective of creating human life for the sole purpose of preparing therapeutic material would clearly not be the dignity of the life created.

Analyzing the use of cloning as a means of combating sterility is much more difficult, as the explicit goal is to create a life with the right to dignity. Moreover, individuals are in no way determined entirely by their genome, as of course the family, cultural and social environment have a powerful "humanizing" influence on the construction of a personality. Two human clones born several decades apart would be much more different psychologically than two identical twins raised in the same family.

Nonetheless, part of the individuality and dignity of a person probably lies in the uniqueness and the unpredictability surrounding his or her development. As a result, the uncertainty of the great lottery of heredity constitutes the principal protection for human beings against biological predetermination imposed by third parties, including parents. One of the blessings of the relationship between parents and their children is their inevitable difference, which results in parents loving their children for what they are, rather than endeavoring to make them what they want. Allowing cloning to circumvent sterility would inevitably lead to it being tolerated in other cases where it was imposed, for example, by authorities.

What would the world be like if we accepted that human "creators" could assume the right to generate creatures in their own likeness, beings whose every biological characteristics would be subjugated to an outside will, copies of bodies that have already lived, half slaves, half fantasies of immortalities?

The results of Wilmut et al. undoubtedly have much merit. One effect of them is to oblige us to face up to our responsibilities. It is not a technical barrier that will protect us from the perspectives I have mentioned, but a moral one, originating from a reflection of the basis of our dignity. That barrier is certainly the most dignified aspect of human genius.

WILL CLONING HARM
PEOPLE?

GREGORY E. PENCE

THE MOST IMPORTANT MORAL OBJECTION TO ORIGINATING A HUMAN by cloning is the claim that the resulting person may be unnecessarily harmed, either by something in the process of cloning or by the unique expectations placed upon the resulting child. This essay considers this kind of objection.

By now the word "cloning" has so many bad associations from science fiction and political demagoguery that there is no longer any good reason to continue to use it. A more neutral phrase, meaning the same thing, is "somatic cell nuclear transfer" (SCNT), which refers to the process by which the genotype of an adult, differentiated cell can be used to create a new human embryo by transferring its nucleus to an enucleated human egg. The resulting embryo can then be gestated to create a baby who will be a delayed twin of its genetic ancestor.

For purposes of clarity and focus, I will only discuss the simple case where a couple wants to originate a single child by SCNT and not the cases of multiple origination of the same genotype. I will also not discuss questions of who would regulate reproduction of genotypes and processes of getting consent to reproduce genotypes.

PARALLELS WITH IN VITRO FERTILIZATION: REPEATING HISTORY?

Any time a new method of human reproduction may occur, critics try to prevent it by citing possible harm to children. The implicit premise: before it is allowed, any new method must prove that only healthy children will be created. Without such proof, the new method amounts to "unconsented to" experimentation on the unborn. So argued the late conservative, Christian bioethicist Paul Ramsey in the early 1970s about in vitro fertilization (IVF).[1]

Of course, ordinary sexual reproduction does not guarantee healthy children every time. Nor can a person consent until he is born. Nor can he really consent until he is old enough to understand consent. The requirement of "consent to be born" is silly.

Jeremy Rifkin, another critic of IVF in the early 1970s, seemed to demand that new forms of human reproduction be risk-free.[2] Twenty years later, Rifkin predictably bolted out the gate to condemn human cloning, demanding its world-wide ban, with penalties for transgressions as severe as those for rape and murder: "It's a horrendous crime to make a Xerox of someone," he declared ominously. "You're putting a human into a genetic straitjacket. For the first time, we've taken the principles of industrial design—quality control, predictability—and applied them to a human being."[3]

Daniel Callahan, a philosopher who had worked in the Catholic tradition and who founded the Hastings Center for research in medical ethics, argued in 1978 that the first case of IVF was "probably unethical" because there was no possible guarantee that Louise Brown would be normal.[4] Callahan added that many medical breakthroughs are unethical because we cannot know (using the philosopher's strong sense of "know") that the first patient will not be harmed. Two decades later, he implied that human cloning would also be unethical: "We live in a culture that likes science and technology very much. If someone wants something, and the rest of us can't prove they are going to do devastating harm, they are going to do it."[5]

Leon Kass, a social conservative and biologist-turned-bioethicist, argued strenuously in 1971 that babies created by artificial fertilization might be deformed: "It doesn't matter how many times the baby is tested while in the mother's womb," he averred, "they will never be certain the baby won't be born without defect."[6]

What these critics overlooked is that no reasonable approach to life avoids all risks. Nothing in life is risk-free, including having children. Even if babies are born healthy, they do not always turn out as hoped. Taking such chances is part of becoming a parent.

Without some risk, there is no progress, no advance. Without risk, pioneers don't cross prairies, astronauts don't walk on the moon, and Freedom Riders don't take buses to integrate the South. The past critics of assisted reproduction demonstrated a psychologically normal but nevertheless unreasonable tendency to magnify the risk of a harmful but unlikely result. Such a result—even if very bad—still represents a very small risk. A baby born with a lethal genetic disease is an extremely bad but unlikely result; nevertheless, the risk shouldn't deter people from having children.

HUMANITY WILL NOT BE HARMED

Human SCNT is even more new and strange-sounding than in vitro fertilization (IVF). All that means is that it will take longer to get used to. Scaremongers have predicted terrible harm if children are born by SCNT, but in fact very little will change. Why is that?

First, to create a child by SCNT, a couple must use IVF, which is an expensive process, costing about $8,000 per attempt. Most American states do not require insurance companies to cover IVF, so IVF is mostly a cash-and-carry operation. Second, most IVF attempts are unsuccessful. The chances of any couple taking home a baby is quite low—only about 15%.

Only about 40,000 IVF babies have been born in America since the early 1980s. Suppose 50,000 such babies are born over the next decade. How many of these couples would want to originate a child by SCNT? Very few—at most, perhaps, a few hundred.

These figures are important because they tamp down many fears. As things now stand, originating humans by SCNT will never be common. Neither evolution nor old-fashioned human sex is in any way threatened. Nor is the family or human society. Most fears about human cloning stem from ignorance.

Similar fears linking cloning to dictatorship or the subjugation of women are equally ignorant. There are no artificial wombs (predictions, yes; realities, no—otherwise we could save premature babies born before 20 weeks). A healthy woman must agree to gestate any SCNT baby and such a woman will retain her right to abort. Women's rights to abortion are checks on evil uses of any new reproductive technology.

NEW THINGS MAKE US FEAR HARMS IRRATIONALLY

SCNT isn't really so new or different. Consider some cases on a continuum. In the first, the human embryo naturally splits in the process of twinning and produces two genetically-identical twins. Mothers have been conceiving and gestating human twins for all of human history. Call the children who result from this process Rebecca and Susan.

In the second case a technique is used where a human embryo is deliberately twinned in order to create more embryos for implantation in a woman who has been infertile with her mate. Instead of a random quirk in the uterus, now a physician and an infertile couple use a tiny electric current to split the embryo. Two identical embryos are created. All embryos are implanted and, as sometimes happens, rather than no embryo implanting successfully or only one, both embryos implant. Again, Rebecca and Susan are born.

In the third case, one of the twinned embryos is frozen (Susan) along with other embryos from the couple and the other embryo is implanted. In this case, although several embryos were implanted, only the one destined to be Rebecca is successful. Again, Rebecca is born.

Two years pass, and the couple desires another child. Some of their frozen embryos are thawed and implanted in the mother. The couple knows that one of the implanted embryos is the twin of Rebecca. In this second round of reproductive assistance, the embryo destined to be Susan successfully implants and a twin is born. Now Susan and Rebecca exist as twins, but born two years apart. Susan is the delayed twin of Rebecca. (Rumors abound that such births have already occurred in American infertility clinics.)

Suppose now that the "embryo that could become Susan" was twinned, and the "non-Susan" embryo is frozen. The rest of the details are then the same as the last scenario, but now two more years pass and the previously-frozen embryo is now implanted, gestated, and born. Susan and Rebecca now have another identical sister, Samantha. They would be identical triplets, born two and four years apart. In contrast to SCNT, where the mother's contribution of mitochondrial genes introduces small variations in nearly-identical genotypes, these embryos would have identical genomes.

Next, suppose that the embryo that could have been Rebecca miscarried and never became a child. The twinned embryo that could become Susan still exists. So the parents implant this embryo and Susan is born. Query to National Bioethics Advisory Commission: have the parents done something illegal? A child has been born who was originated by reproducing an embryo with a unique genotype. Remember, the embryo-that-could-become Rebecca existed first. So Susan only exists as a "clone" of the non-existent Rebecca.

Now, as bioethicist Leroy Walters emphasizes, let us consider an even thornier but more probable scenario.[7] Suppose we took the embryo-that-could-become Susan and transferred its nucleus to an enucleated egg of Susan's mother. Call the person who will emerge from this embryo "Suzette," because she is like Susan but different, because of her new mitochondrial DNA. Although the "Susan" embryo was created sexually, Suzette's origins are through somatic cell nuclear transfer. It is not clear that this process is illegal. The NBAC *Report* avoids taking a stand on this kind of case.[8]

Now compare all the above cases to originating Susan asexually by SCNT from the genotype of the adult Rebecca. Susan would again have a nearly-identical genome with Rebecca (identical except for mitochondrial DNA contributed by the gestating woman). Here we have nearly identical female genotypes, separated in time, created by choice. But how is this so different from choosing to have a delayed twin-child? Originating a child by SCNT is not a breakthrough in kind but a matter of degree along a continuum involving twins and a special kind of reproductive choice.

COMPARING THE HARMS OF HUMAN REPRODUCTION

The question of multiple copies of one genome and its special issues of harm are ones that will not be discussed in this essay, but one asymmetry in our moral intuitions should be noticed.

The increasing use of fertility drugs has expanded many times the number of humans born who are twins, triplets, quadruplets, quintuplets, sextuplets, and even (in November of 1997 to the McCaugheys of Iowa) septuplets. If an entire country can rejoice about seven humans who are gestated in the same womb, raised by the same parents, and simultaneously created randomly from the same two sets of chromosomes, why should the same country fear deliberately originating copies of the same genome, either at once or over time? Our intuitions are even more skewed when we rejoice in the statistically-unlikely case of the seven healthy McCaughey children and ignore the far more likely cases where several of the multiply-gestated fetuses are disabled or dead.

People exaggerate the fears of the unknown and downplay the very real dangers of the familiar. In a very important sense, driving a car each day is far more dangerous to children than the new form of human reproduction under discussion here. Many, many people are hurt and killed every day in automobile wrecks, yet few people consider not driving.

In SCNT, there are possible dangers of telomere shortening, inheritance of environmental effects on adult cells passed to embryonic cells, and possible unknown dangers. Mammalian animal studies must determine if such dangers will occur in human SCNT origination. Once such studies prove that there are no special dangers of SCNT, the crucial question will arise: how safe must we expect human SCNT to be before we allow it?

In answering this question, it is very important to ask about the baseline of comparison. How safe is ordinary, human sexual reproduction? How safe is assisted reproduction? Who or what counts as a subject of a safety calculation about SCNT?

At least 40% of human embryos fail to implant in normal sexual reproduction.[9] Although this fact is not widely known, it is important because some discussions tend to assume that every human embryo becomes a human baby unless some extraordinary event occurs such as abortion. But this is not true. Nature seems to have a genetic filter, such that malformed embryos do not implant. About 50% of the rejected embryos are chromosomally abnormal, meaning that if they were somehow brought to term, the resulting children would be mutants or suffer genetic dysfunction.

A widely-reported but misleading aspect of Ian Wilmut's work was that it took 277 embryos to produce one live lamb. In fact, Wilmut started with 277 eggs, fused nuclei with them to create embryos, and then allowed them to become the best 29 embryos, which were allowed to gestate further. He had three lambs almost live, with one true success, Dolly. Subsequent work may easily bring the efficiency rate to 25%. When the calves "Charlie" and "George" were born in 1998, four live-born calves were created from an initial batch of only 50 embryos.[10]

Wilmut's embryo-to-birth ratio only seems inefficient or unsafe because the real inefficiency fate of accepted forms of human assisted reproduction is so little known. In in vitro fertilization, a woman is given drugs to stimulate super-ovulation so that physicians can remove as many eggs as possible. At each cycle of attempted in vitro fertilization, three or four embryos are implanted. Most couples make several attempts, so as many as nine to twelve embryos are involved for each couple. As noted, only about 15–20% of couples undergoing such attempts ever take home a baby.

Consider what these numbers mean when writ large. Take a hundred couples attempting assisted reproduction, each undergoing (on average) three attempts. Suppose there are unusually good results and that 20% of these couples eventually take home a baby. Because more than one embryo may implant, assume that among these 20 couples, half have non-identical twins. But what is the efficiency rate here? Assuming a low number of three embryos implanted each time for the 300 attempts, it will take 900 embryos to produce 30 babies, for an efficiency rate of 1 in 30.

Nor is it true that all the loss of human potential occurred at the embryonic stage. Unfortunately, some of these pregnancies will end in miscarriages of fetuses, some well along in the second trimester.

Nevertheless, such loss of embryos and fetuses is almost universally accepted as morally permissible. Why is that? Because the infertile parents are trying to conceive their own children, because everyone thinks that is a good motive, and because few people object to the loss of embryos and fetuses *in this context of trying to conceive babies*. Seen in this light, what Wilmut did, starting out with a large number of embryos to get one successful lamb at birth, is not so novel or different from what now occurs in human assisted reproduction.

SUBJECTS AND NONSUBJECTS OF HARM

One premise that seems to figure in discussions of the safety of SCNT and other forms of assisted reproduction is that loss of human embryos morally matters. That premise should be rejected.

As the above discussion shows, loss of human embryos is a normal part of human conception and, without this process, humanity might suffer much more genetic disease. This process essentially involves the loss of human embryos as part of the natural state of things. Indeed, some researchers believe that for every human baby successfully born, there has been at least one human embryo lost along the way.

In vitro fertilization is widely-accepted as a great success in modern medicine. As said, over 40,000 American babies have been born this way. But calculations indicate that as many as a million human embryos may have been used in creating such successes.

Researchers often create embryos for subsequent cycles of implantation, only to learn that a pregnancy has been achieved and that such stored embryos are no longer needed. Thousands of such embryos can be stored indefinitely in liquid nitrogen. No one feels any great urgency about them and, indeed, many couples decline to pay fees to preserve their embryos.

The above considerations point to the obvious philosophical point that embryos are not persons with rights to life. Like an acorn, their value is all potential, little actual. Faced with a choice between paying a thousand dollars to keep two thousand embryos alive for a year in storage, or paying for an operation to keep a family pet alive for another year, no one will choose to pay for the embryos. How people actually act says much about their real values.

Thus an embryo cannot be harmed by being brought into existence and then being taken out of existence. An embryo is generally considered such until nine weeks after conception, when it is called a "fetus" (when it is born, it is called a "baby"). Embryos are not sentient and cannot experience pain. They are thus not the kind of subjects that can be harmed or benefitted.

As such, whether it takes one embryo to create a human baby or a hundred does not matter morally. It may matter aesthetically, financially, emotionally, or in time spent trying to reproduce, but it does not matter morally. As such, new forms of human reproduction such as IVF and SCNT that involve significant loss of embryos cannot be morally criticized on this charge.

Finally, because embryos don't count morally, they could be tested in various ways to eliminate defects in development or genetic mishaps. Certainly, if four or five SCNT embryos were implanted, only the healthiest one should be brought to term. As such, the risk of abnormal SCNT babies could be minimized.

SETTING THE STANDARD ABOUT THE RISK OF HARM

Animal tests have not yet shown that SCNT is safe enough to try in humans, and extensive animal testing should be done over the next few years. That means that, before we attempt SCNT in humans, we will need to be able to routinely produce healthy offspring by SCNT in lambs, cattle, and especially, non-human primates. After this testing is done, the time will come when a crucial question must be answered: how safe must human SCNT be before it is allowed? This is probably the most important, practical question before us now.

Should we have a very high standard, such that we take virtually no risk with a SCNT child? Daniel Callahan and Paul Ramsey, past critics of IVF, implied that unless a healthy baby could be guaranteed the first time, it was unethical to try to produce babies in a new way. At the other extreme, a low standard would allow great risks.

What is the appropriate standard? How high should be the bar over which scientists must be made to jump before they are allowed to try to originate a SCNT child? In my opinion, the standard of Callahan and Ramsey is too high. In reality, only God can meet that Olympian standard. It is also too high for those physicians trying to help infertile couples. If this high standard had been imposed on these people in the past, no form of assisted reproduction—including in vitro fertilization—would ever have been allowed.

On the other end of the scale, one could look at the very worst conditions for human gestation, where mothers are drug-dependent during pregnancy or exposed to dangerous chemicals. Such worst-case conditions include parents with a 50% chance of passing on a lethal genetic disease. The lowest standard of harm allows human reproduction even if there is such a high risk of harm ("harm" in the sense that the child would likely have a sub-normal future). One could argue that since society allows such mothers and couples to reproduce sexually, it could do no worse by allowing a child to be originated by SCNT.

I believe that the low standard is inappropriate to use with human SCNT. There is no reason to justify down to the very worst conditions under which society now tolerates humans being born. If the best we can do by SCNT is to produce children as good as those born with fetal-maternal alcohol syndrome, we shouldn't originate children this way.

Between these standards, there is the normal range of risk that is accepted by ordinary people in sexual reproduction. Human SCNT should be allowed when the predicted risk from animal studies falls within this range. "Ordinary people" refers to those who are neither alcoholic nor dependent on an illegal drug and where neither member of the couple knowingly passes on a high risk for a serious genetic disease.

This standard seems reasonable. It does not require a guarantee of a perfect baby, but it also rejects the "anything goes" view. For example, if the rate of serious deformities in normal human reproduction is 1%, and if the rate of chimpanzee SCNT reproduction were brought down to this rate, and if there were no reason to think that SCNT in human primates would be any higher, it should be permissible to attempt human SCNT.

WE ALREADY ALLOW MORE DANGEROUS FORMS OF FAMILIAR THINGS

Consider the case of an infertility researcher trying to increase the chances of older women successfully giving birth by using eggs of younger women. The practice of using an egg of a younger woman increases the chances of a 44-year-old woman giving birth from the 3.5% chance of IVF to 50%.[11] The problem with this practice is that the gestating, older woman does not have a genetic connection to the resulting child.

But there is a new, possible way to solve this problem. Researcher James Grifo, chief of New York University's infertility clinic, proposes to enucleate an egg of a younger women and insert the sexually-mixed chromosomes into it from an embryo of an older couple.[12] The older woman will gestate the new SCNT-created embryo, an embryo that may successfully implant because it has the outer cytoplasm and mitochondria of a younger woman's egg. Creating an embryo by such a SCNT process is not "cloning," not asexual reproduction, because the chromosomes randomly mixed when sperm met egg. Nevertheless, the process of creating a human embryo (created sexually) by SCNT here is obviously very close to the process of creating a human embryo by *asexual* SCNT.

This new, nucleus-transferring procedure is controversial but permitted under existing law. It undoubtedly is experimental and risks possible harm to any resulting child, but because this procedure does not have the emotional associations of "cloning," it will probably pass unnoticed in the general world. This is exactly what happened with intracytoplasmic sperm injection—where only one sperm is used to fertilize an egg—despite its unknown safety record when first attempted.[13] Such new procedures, and the possible harm of doing them, are very close to those of SCNT. We say the sky will fall if we try SCNT while we ignore the fact that very similar risks are being taken all around us.

Finally, we are already doing things far more radical than human SCNT. Putting human genes in pigs to create possible organ transplants from such altered pigs is far more radical than human SCNT. Transplants from such pig organs open up the possibility of a two-way travel of porcine viruses to humans and vice-versa (of concern to those who think that AIDS came from simian–human contact). In 1987, we allowed Harvard University to patent its oncomouse (aka the "Harvard oncomouse"). Since then over a thousand applications for such patents have been filed and over 50 patents on genetically-altered or genetically-created animals have been issued by the U.S. Patent Office.[14] Overall, from genetically-altered tomatoes to pig-grown livers for transplanting into humans, we are doing radically new things to save human lives, crossing natural barriers all the time, and hardly blinking an eye about it. Why, then, are we so concerned about SCNT?

PSYCHOLOGICAL HARM TO THE CHILD

Another concern is about psychological harm to a child originated by SCNT. According to this objection, choosing to have a child is not like choosing a car or house. It is a moral decision because another being is affected. Having a child should be a careful, responsible choice and focused on what's best for the child. Having a child originated by SCNT is not morally permissible because it is not best for the child.

The problem with this argument is the last six words of the last sentence, which assumes bad motives on the part of parents. Unfortunately, SCNT is associated with bad motives in science fiction, but until we have evidence that it will be used this way, why assume the worst about people?

Certainly, if someone deliberately brought a child into the world with the intention of causing him harm, that would be immoral. Unfortunately, the concept of harm is a continuum and some people have very high standards, such that not providing a child a stay-at-home parent constitutes harming the child. But there is nothing about SCNT per se that is necessarily linked to bad motives. True, people would have certain expectations of a child created by SCNT, but parents-to-be already have certain expectations about children.

Too many parents are fatalistic and just accept whatever life throws at them. The very fact of being a parent for many people is something they must accept (because abortion was not a real option). Part of this acceptance is to just accept whatever genetic combination comes at birth from the random assortment of genes.

But why is such acceptance a good thing? It is a defeatist attitude in medicine against disease; it is a defeatist attitude toward survival when one's culture or country is under attack; and it is a defeatist attitude toward life in general. "The expectations of parents will be too high!" critics repeat. "Better to leave parents in ignorance and to leave their children as randomness decrees." The silliness of that view is apparent as soon as it is made explicit.

If we are thinking about harm to the child, an objection that comes up repeatedly might be called the argument for an open future. "In the case of cloning," it is objected, "the expectations are very specifically tied to the life of another person. So in a sense, the child's future is denied to him because he will be expected to be like his ancestor. But part of the wonder of having children is surprise at how they turn out. As such, some indeterminacy should remain a part of childhood. Human SCNT deprives a person of an open future because when we know how his previous twin lived, we will know how the new child will live."

It is true that the adults choosing this genotype rather than that one must have some expectations. There has to be some reason for choosing one genotype over another. But these expectations are only half based in fact. As we know, no person originated by SCNT will be identical to his ancestor because of mitochondrial DNA, because of his different gestation, because of his different parents, because of his different time in history, and perhaps, because of his different country and culture. Several famous pairs of conjoined twins, such as Eng and Chang, with both identical genotypes and identical uterine/child-hood environments, have still had different personalities.[15] To assume that a SCNT child's future is not open is to assume genetic reductionism.

Moreover, insofar as parents have specific expectations about children created by SCNT, such expectations will likely be no better or worse than the normal expectations by parents of children created sexually. As said, there is nothing about SCNT per se that necessitates bad motives on the part of parents.

Notice that most of the expected harm to the child stems *from the predicted, prejudicial attitudes of other people to the SCNT child*. ("Would you want to be a cloned child? Can you imagine being called a freak and having only one genetic parent?") As such, it is important to remember that social expectations are *merely* social expectations. They are malleable and can change quickly. True, parents might initially have expectations that are too high and other people might regard such children with prejudice. But just as such inappropriate atti-tudes faded after the first cases of in vitro fertilization, they will fade here too.

Ron James, the Scottish millionaire who funded much of Ian Wilmut's re-search, points out that social attitudes change fast. Before the announcement of Dolly, polls showed that people thought that cloning animals and gene trans-fer to animals were "morally problematic," whereas germ-line gene therapy fell in the category of "just wrong." Two months after the announcement of Dolly, and after much discussion of human cloning, people's attitudes had shifted to accepting animal cloning and gene transfer to humans as "morally permissible," whereas germ-line gene therapy had shifted to being merely "morally problematic."[16]

James Watson, the co-discoverer of the double helix, once opposed in vitro fertilization by claiming that prejudicial attitudes of other people would harm children created this way (see this volume).[17] In that piece, the prejudice was really in Watson, because the way that he was stirring up fear was doing more to create the prejudice than any normal human reaction. Similarly, Leon Kass's recent long essay in *The New Republic* (see this volume), where he calls human asexual reproduction "repugnant" and a "horror," creates exactly the kind of prejudiced reaction that he predicts.[18] Rather than make a priori, self-fulfilling prophecies, wouldn't it be better to be empirical about such matters? To be more optimistic about the reactions of ordinary parents?

Children created by SCNT would not *look* any different from other chil-dren. Nobody at age two looks like he does at age 45 and, except for his parents, nobody knows what the 45-year-old man looked liked at age two. And since ordinary children often look like their parents, know one would be able to tell a SCNT child from others until he had lived a decade.

Kass claims that a child originated by SCNT will have "a troubled psychic identity" because he or she will be "utterly" confused about his social, genetic, and kinship ties.[19] At worst, this child will be like a child of "incest" and may, if originated as a male from the father, have the same sexual feelings towards the wife as the father. An older male might in turn have strong sexual feelings toward a young female with his wife's genome.

Yet if this were so, any husband of any married twin might have an equally troubled psychic identity because he might have the same sexual feelings toward the twin as his wife. Instead, those in relationships with twins claim that the individuals are very different.

Much of the above line of criticism simply begs the question and assumes that humans created by SCNT will be greeted by stigma or experience confusion. It is hard to understand why, once one gets beyond the novelty, because a child created asexually would know *exactly* who his ancestor was. No confusion there. True, prejudicial expectations could damage children, but why make public policy based on that?

Besides, isn't this kind of argument hypocritical in our present society? Where no one is making any serious effort to ban divorce, despite the overwhelming evidence that divorce seriously damages children, even teenage children. It is always far easier to concentrate on the dramatic, far-off harm than the ones close-at-hand. When we are really concerned about minimizing harm to children, we will pass laws requiring all parents wanting to divorce to go through counseling sessions or to wait a year. We will pass a federal law compelling child-support from fathers who flee to other states, and make it impossible to renew a professional license or get paid in a public institution in another state until all child-support is paid. After that is done, then we can non-hypocritically talk about how much our society cares about not harming children who may be originated in new ways.

In conclusion, the predicted harms of SCNT to humans are wildly exaggerated, lack a comparative baseline, stem from irrational fears of the unknown, overlook greater dangers of familiar things, and are often based on the armchair psychological speculation of amateurs. Once studies prove SCNT as safe as normal sexual reproduction in non-human mammals, the harm objection will disappear. Given other arguments that SCNT could substantially benefit many children, the argument that SCNT would harm children is a weak one that needs to be weighed against its many potential benefits.[20]

---

1. Paul Ramsey, *Fabricated Man: The Ethics of Genetic Control* (New Haven, Conn.: Yale University Press, 1970).

2. "What are the psychological implications of growing up as a specimen, sheltered not by a warm womb but by steel and glass, belonging to no one but the lab technician who joined together sperm and egg? In a world already populated with people

with identity crises, what's the personal identity of a test-tube baby?" J. Rifkin and T. Howard, *Who Shall Play God?* (New York: Dell, 1977), 15.

3. Ehsan Massod, "Cloning Technique 'Reveals Legal Loophole'," *Nature* 38, 27 February 1987.

4. *New York Times*, 27 July 1978, A16.

5. Knight-Ridder newspapers, 10 March 1997.

6. Leon Kass, "The New Biology: What Price Relieving Man's Estate?" *Journal of the American Medical Association*, vol. 174, 19 November 1971, 779–788.

7. Leroy Walters, "Biomedical Ethics and Their Role in Mammalian Cloning," Conference on Mammalian Cloning: Implications for Science and Society, 27 June 1997, Crystal City Marriott, Crystal City, Virginia.

8. National Bioethics Advisory Commission (NBAC), *Cloning Human Beings: Report and Recommendations of the National Bioethics Advisory Commission*, Rockville, Md., June 1997.

9. A. Wilcox et al., "Incidence of Early Loss of Pregnancy," *New England Journal of Medicine* 319, no. 4, 28 July 1988, 189–194. See also J. Grudzinskas and A. Nysenbaum, "Failure of Human Pregnancy after Implantation," *Annals of New York Academy of Sciences* 442, 1985, 39–44; J. Muller et al., "Fetal Loss after Implantation," *Lancet* 2, 1980, 554–556.

10. Rick Weiss, "Genetically Engineered Calves Cloned," 21 January 1998, *Washington Post*, A3.

11. Lisa Belkin, "Pregnant with Complications," *New York Times Magazine*, October 26, 1997, 38.

12. ABC News report, October 27, 1997.

13. Axel Kahn, "Clone Animals . . . Clone Man?" specially-commissioned article to accompany articles from *Nature* on cloning on the web site of *Nature* (reprinted in this volume).

14. See the web site of a leading law firm in this area, Elman & Associates, at http://www.elman.com/elman.

15. David R. Collins, *Eng and Chang: The Original Siamese Twins* (New York: Dillon Press, 1994). Elaine Landau, *Joined at Birth: The Lives of Conjoined Twins* (New York: Grolier Publishing, 1997). See also Geoffrey A. Machin, "Conjoined Twins: Implications for Blastogenesis," *Birth Defects: Original Articles Series* 20, no. 1, 1993, March of Dimes Foundation, 142.

16. Ron James, Managing Director, PPL Therapeutics, "Industry Perspective: The Promise and Practical Applications," Conference on Mammalian Cloning: Implications for Science and Society, 27 June 1997, Crystal City Marriott, Crystal City, Virginia.

17. James D. Watson, "Moving Towards the Clonal Man," *Atlantic*, May 1971, 50–53.

18. Leon Kass, "The Wisdom of Repugnance," *The New Republic*, 2 June 1997.

19. Kass, "The Wisdom of Repugnance," 22–23.

20. Thanks to Mary Litch for comments on this essay.

THE CONFUSION

OVER CLONING

R . C . LEWONTIN

R. C. Lewontin wrote The Genetic Basis of Evolutionary Change *(1974), a classic in the field of population genetics, which emphasizes that there is more genetic variation within races than between them. He wrote* Biology and Identity *and cowrote* Not in Our Genes *(1984). A fierce socialist critic of reductionist views in biology, genetics, intelligence, and politics, Lewontin most typically argues that observed differences between different racial/ethnic groups are primarily of nonbiological origin. He is the Alexander Agassiz Professor of Zoology and Biology at Harvard University.*

In this article, Lewontin argues that our knowledge of identical human twins should lead us not to fear humans originated from the same genotype. He also attacks the simplistic notion that "person" is the same as "genes" and skewers NBAC for lapsing into such simplistic views. He is especially critical of the attention that NBAC gave to "religious perspectives" on human cloning in contrast to "ethical perspectives."

THERE IS NOTHING LIKE SEX OR VIOLENCE FOR CAPTURING THE IM-mediate attention of the state. Only a day after Franklin Roosevelt was told in October 1939 that both German and American scientists could probably make an atom bomb, a small group met at the President's direction to talk about the problem and within ten days a committee was undertaking a full-scale investigation of the possibility. Just a day after the public announcement on February 23, 1997, that a sheep, genetically identical to another sheep, had been produced by cloning, Bill Clinton formally requested that the National Bioethics Advisory Commission "undertake a thorough review of the legal and ethical issues associated with the use of this technology. . . ."

The President had announced his intention to create an advisory group on bioethics eighteen months before, on the day that he received the disturbing report of the cavalier way in which ionizing radiation had been administered experimentally to unsuspecting subjects.[1] The commission was finally formed, after a ten-month delay, with Harold Shapiro, President of Princeton, as chair and a membership consisting largely of academics from the fields of philosophy, medicine, public health, and law, a representation from government and private foundations, and the chief business officer of a pharmaceutical company. In his letter to the commission the President referred to "serious ethical questions, particularly with respect to the possible use of this technology to clone human embryos" and asked for a report within ninety days. The commission missed its deadline by only two weeks.

In order not to allow a Democratic administration sole credit for grappling with the preeminent ethical issue of the day, the Senate held a day-long inquiry on March 12, a mere three weeks after the announcement of Dolly. Lacking a body responsible for any moral issues outside the hanky-panky of its own membership, the Senate assigned the work to the Subcommittee on Public Health and Safety of the Committee on Labor and Human Resources, perhaps on the grounds that cloning is a form of the production of human resources. The testimony before the subcommittee was concerned not with issues of the health and safety of labor but with the same ethical and moral concerns that preoccupied the bioethics commission. The witnesses representing the biotechnology industry were especially careful to assure the senators that they would not dream of making whole babies and were interested in cloning solely as a laboratory method for producing cells and tissues that could be used in transplantation therapies.

It seems pretty obvious why, just after the Germans' instant success in Poland, Roosevelt was in a hurry. The problem, as he said to Alexander Sachs, who first informed him about the possibility of the Bomb, was to "see that the Nazis don't blow us up." The origin of Mr. Clinton's sense of urgency is not so clear. After all, it is not as if human genetic clones don't appear every day of the week, about thirty a day in the United States alone, given that there are about four million births a year with a frequency of identical twins of roughly 1 in 400.[2] So it cannot be the mere existence of dopplegänger that creates urgent problems (although I will argue that parents of twins are often guilty of a kind of psychic child abuse). And why ask the commission on bioethics rather than a technical committee of the National Institutes of Health or the National Research Council? Questions of individual autonomy and responsibility for one's own actions, of the degree to which the state ought to interpose itself in matters of personal decision, are all central to the struggle over smoking, yet the bioethics commission has not been asked to look into the bioethics of tobacco, a matter that would certainly be included in its original purpose.

The answer is that the possibility of human cloning has produced a nearly universal anxiety over the consequences of hubris. The testimony before the bioethics commission speaks over and over of the consequences of "playing God." We have no responsibility for the chance birth of genetically identical individuals, but their deliberate manufacture puts us in the Creation business, which, like extravagant sex, is both seductive and frightening. Even Jehovah botched the job despite the considerable knowledge of biology that He must have possessed, and we have suffered the catastrophic consequences ever since. According to Haggadic legend, the Celestial Cloner put a great deal of thought into technique. In deciding on which of Adam's organs to use for Eve, He had the problem of finding tissue that was what the biologist calls "totipotent," that is, not already committed in development to a particular function. So He cloned Eve

> not from the head, lest she carry her head high in arrogant pride, not from the eye, lest she be wanton-eyed, not from the ear lest she be an eavesdropper, not from the neck lest she be insolent, not from the mouth lest she be a tattler, not from the heart lest she be inclined to envy, not from the hand lest she be a meddler, not from the foot lest she be a gadabout

but from the rib, a "chaste portion of the body." In spite of all the care and knowledge, something went wrong, and we have been earning a living by the sweat of our brows ever since. Even in the unbeliever, who has no fear of sacrilege, the myth of the uncontrollable power of creation has a resonance that gives us all pause. It is impossible to understand the incoherent and unpersuasive document produced by the National Bioethics Advisory Commission except as an attempt to rationalize a deep cultural prejudice, but it is also impossible to understand it without taking account of the pervasive error that confuses the genetic state of an organism with its total physical and psychic nature as a human being.

After an introductory chapter placing the issue of cloning in a general historical and social perspective, the commission begins with an exposition of the technical details of cloning and with speculations on the reproductive, medical, and commercial applications that are likely to be found for the technique. Some of these applications involve the clonal reproduction of genetically engineered laboratory animals for research or the wholesale propagation of commercially desirable livestock; but these raised no ethical issues for the commission, which, wisely, avoided questions of animal rights.

Specifically human ethical questions are raised by two possible applications of cloning. First, there are circumstances in which parents may want to use techniques of assisted reproduction to produce children with a known genetic makeup for reasons of sentiment or vanity or to serve practical ends. Second, there is the possibility of producing embryos of known genetic constitution whose cells and tissues will be useful for therapeutic purposes. Putting aside, for consideration in a separate chapter, religious claims that human cloning violates various scriptural and doctrinal prescriptions about the correct relation between God and man, men and women, husbands and wives, parents and children, or sex and reproduction, the commission then lists four ethical issues to be considered: individuality and autonomy, family integrity, treating children as objects, and safety.

The most striking confusion in the report is in the discussion of individuality and autonomy. Both the commission report and witnesses before the Senate subcommittee were at pains to point out that identical genes do not make identical people. The fallacy of genetic determinism is to suppose that the genes "make" the organism. It is a basic principle of developmental biology that organisms undergo a continuous development from conception to death, a development that is the unique consequence of the interaction of the genes in their cells, the temporal sequence of environments through which the organisms pass, and random cellular processes that determine the life, death, and transformations of cells. As a result, even the fingerprints of identical twins are not identical. Their temperaments, mental processes, abilities, life choices, disease histories, and deaths certainly differ despite the determined efforts of many parents to enforce as great a similarity as possible.

Frequently twins are given names with the same initial letter, dressed identically with identical hair arrangements, and given the same books, toys, and training. There are twin conventions at which prizes are offered for the most similar pairs. While identical genes do indeed contribute to a similarity between them, it is the pathological compulsion of their parents to create an inhuman identity between them that is most threatening to the individuality of genetically identical individuals.

But even the most extreme efforts to turn genetic clones into human clones fail. As a child I could not go to the movies or look at a picture magazine without being confronted by the genetically identical Dionne quintuplets, identically dressed and coiffed, on display in "Quintland" by Dr. Dafoe and the Province of Ontario for the amusement of tourists. This enforced homogenization continued through their adolescence, when they were returned to their parents' custody. Yet each of their unhappy adulthoods was unhappy in its own way, and they seemed no more alike in career or health than we might expect from five girls of the same age brought up in a rural working-class

French Canadian family. Three married and had families. Two trained as nurses, two went to college. Three were attracted to a religious vocation, but only one made it a career. One died in a convent at age twenty, suffering from epilepsy, one at age thirty-six, and three remain alive at sixty-three. So much for the doppelgänger phenomenon. The notion of "cloning Einstein" is a biological absurdity.

The Bioethics Advisory Commission is well aware of the error of genetic determinism, and the report devotes several pages to a sensible and nuanced discussion of the difference between genetic and personal identity. Yet it continues to insist on the question of whether cloning violates an individual human being's "unique qualitative identity."

> And even if it is a mistake to believe such crude genetic determinism according to which one's genes determine one's fate, what is important for oneself is whether one *thinks* one's future is open and undetermined, and so still to be largely determined by one's own choices. [p. A8, emphasis added]

Moreover, the problem of self-perception may be worse for a person cloned from an adult than it is for identical twins, because the already fully formed and defined adult presents an irresistible persistent model for the developing child. Certainly for the general public the belief is widely expressed that a unique problem of identity is raised by cloning that is not already present for twins. The question posed by the commission, then, is not whether genetic identity per se destroys individuality, but whether the erroneous state of public understanding of biology will undermine an individual's own sense of uniqueness and autonomy.

Of course it will, but surely the commission has chosen the wrong target of concern. If the widespread genomania propagated by the press and by vulgarizers of science produces a false understanding of the dominance that genes have over our lives, then the appropriate response of the state is not to ban cloning but to engage in a serious educational campaign to correct the misunderstanding. It is not Dr. Wilmut and Dolly who are a threat to our sense of uniqueness and autonomy, but popularizers like Richard Dawkins who describes us as "gigantic lumbering robots" under the control of our genes that have "created us, body and mind."

Much of the motivation for cloning imagined by the commission rests on the same mistaken synecdoche that substitutes "gene" for "person." In one scenario a self-infatuated parent wants to reproduce his perfection or a single woman wants to exclude any other contribution to her offspring. In another, morally more appealing, story a family suffers an accident that kills the father and leaves an only child on the point of death. The mother, wishing to have a child who is the biological offspring of her dead husband, uses cells from the dying infant to clone a baby. Or what about the sterile man whose entire family has been exterminated in Auschwitz and who wishes to prevent the extinction of his genetic patrimony?

Creating variants of these scenarios is a philosopher's parlor game. All such stories appeal to the same impetus that drives adopted children to search for their "real," i.e., biological, parents in order to discover their own "real" identity. They are modern continuations of an earlier preoccupation with blood as the carrier of an individual's essence and as the mark of legitimacy. It is not the possibility of producing a human being with a copy of someone else's genes that has created the difficulty or that adds a unique element to it. It is the fetishism of "blood" which, once accepted, generates an immense array of apparent moral and ethical problems. Were it not for the belief in blood as essence, much of the motivation for the cloning of humans would disappear.

The cultural pressure to preserve a biological continuity as the form of immortality and family identity is certainly not a human universal. For the Romans, as for the Japanese, the preservation of family interest was the preeminent value, and adoption was a satisfactory substitute for reproduction. Indeed, in Rome the foster child (*alumnus*) was the object of special affection by virtue of having been adopted, i.e., acquired by an act of choice.

The second ethical problem cited by the commission, family integrity, is neither unique to cloning nor does it appear in its most extreme form under those circumstances. The contradictory meanings of "parenthood" were already made manifest by adoption and the old-fashioned form of reproductive technology, artificial insemination from anonymous semen donors. Newer technology like in vitro fertilization and implantation of embryos into surrogate mothers has already raised issues to which the possibility of cloning adds nothing. A witness before the Senate subcommittee suggested that the "replication of a human by cloning would radically alter the definition of a human being by producing the world's first human with a single genetic parent."[3] Putting aside the possible priority of the case documented in Matthew 1:23, there is a confusion here. A child by cloning has a full double set of chromosomes like anyone else, half of which were derived from a mother and half from a father. It happens that these chromosomes were passed through another individual, the cloning donor, on their way to the child. That donor is certainly not the child's "parent" in any biological sense, but simply an earlier offspring of the original parents. Of course this sibling may *claim* parenthood over its delayed twin, but it is not obvious what juridical or ethical principle would impel a court or anyone else to recognize that claim.

There is one circumstance, considered by the commission, in which cloning is a biologically realistic solution to a human agony. Suppose that a child, dying of leukemia, could be saved by a bone marrow replacement. Such transplants are always risky because of immune incompatibilities between the recipient and the donor, and these incompatibilities are a direct consequence of genetic differences. The solution that presents itself is to use bone marrow from a second, genetically identical, child who has been produced by cloning from the first.[4] The risk to a bone marrow donor is not great, but suppose it were a kidney that was needed. There is, moreover, the possibility that the fetus itself is to be sacrificed in order to provide tissue for therapeutic purposes. This scenario presents in its starkest form the third ethical issue of concern to the commission, the objectification of human beings. In the words of the commission:

> To objectify a person is to act towards the person without regard for his or her own desires or well-being, as a thing to be valued according to externally imposed standards, and to control the person rather than to engage her or him in a mutually respectful relationship.

We would all agree that it is morally repugnant to use human beings as mere instruments of our deliberate ends. Or would we? That's what I do when I call in the plumber. The very words "employment" and "employee" are descriptions of an objectified relationship in which human beings are "thing(s) to be valued according to externally imposed standards." None of us escapes the objectification of humans that arises in economic life. Why has no National Commission on Ethics been called into emergency action to discuss the conceptualization of human beings as "factory hands" or "human capital" or "operatives"? The report of the Bioethics Advisory Commission fails to explain how cloning would significantly increase the already immense number of children whose conception and upbringing were intended to make them instruments of their parents' frustrated ambitions, psychic fantasies, desires for immortality, or property calculations.

Nor is there a simple relation between those motivations and the resulting family relations. I myself was conceived out of my father's desire for a male heir, and my mother, not much interested in maternity, was greatly relieved when her first and only child filled the bill. Yet, in retrospect, I am glad they were my parents. To pronounce a ban on human cloning because sometimes it will be used for instrumental purposes misses both the complexity of human motivation and the unpredictability of developing personal relationships. Moreover, cloning does not stand out from other forms of reproductive technology in the degree to which it is an instrument of parental fulfillment. The problem of objectification permeates social relations. By loading all the weight of that sin on the head of one cloned lamb, we neatly avoid considering our own more general responsibility.

The serious ethical problems raised by the prospect of human cloning lie in the fourth domain considered by the bioethics commission, that of safety. Apparently, these problems arise because cloned embryos may not have a proper set of chromosomes. Normally, a sexually reproduced organism contains in all its cells two sets of chromosomes, one received from its mother through the egg and one from the father through the sperm. Each of these sets contains a complete set of the different kinds of genes necessary for normal development and adult function. Even though each set has a complete repertoire of genes, for reasons that are not well understood we must have two sets and only two sets to complete normal development. If one of the chromosomes should accidentally be present in only one copy or in three, development will be severely impaired.

Usually we have exactly two copies in our cells because in the formation of the egg and sperm that combined to produce us, a special form of cell division occurs that puts one and only one copy of each chromosome into each egg and each sperm. Occasionally, however, especially in people in their later reproductive years, this mechanism is faulty and a sperm or egg is produced in which one or another chromosome is absent or present more than once. An embryo conceived from such a faulty gamete will have a missing or extra chromosome. Down's syndrome, for example, results from an extra Chromosome 21, and Edward's syndrome, almost always lethal in the first few weeks of life, is produced by an extra Chromosome 18.

After an egg is fertilized in the usual course of events by a sperm, cell division begins to produce an embryo, and the chromosomes, which were in a resting state in the original sperm and egg, are induced to replicate new copies by signals from the complex machinery of cell division. The division of the cells and the replication of more chromosome copies are in perfect synchrony so every new cell gets a complete exact set of chromosomes just like the fertilized egg. When clonal reproduction is performed, however, the events are quite different. The nucleus containing the egg's chromosomes are removed and the egg cell is fused with a cell containing a nucleus from the donor that already contains a full duplicate set of chromosomes. These chromosomes are not necessarily in the resting state and so they may divide out of synchrony with the embryonic cells. The result will be extra and missing chromosomes so that the embryo will be abnormal and will usually, but not necessarily, die.

The whole trick of successful cloning is to make sure that the chromosomes of the donor are in the right state. However, no one knows how to make sure. Dr. Wilmut and his colleagues know the trick in principle, but they produced only one successful Dolly out of 277 tries. The other 276 embryos died at various stages of development. It seems pretty obvious that the reason the Scottish laboratory did not announce the existence of Dolly until she was a full-grown adult sheep is that they were worried that her postnatal development would go awry. Of course, the technique will get better, but people are not sheep and there is no way to make cloning work reliably in people except to experiment on people. Sheep were chosen by the Scottish group because they had turned out in earlier work to be unusually favorable animals for growing fetuses cloned from embryonic cells. Cows had been tried but without success. Even if the methods could be made eventually to work as well in humans as in sheep, how many human embryos are to be sacrificed, and at what stage of their development?[5] Ninety percent of the loss of the experimental sheep embryos was at the so-called "morula" stage, hardly more than a ball of cells. Of the twenty-nine embryos implanted in maternal uteruses, only one showed up as a fetus after fifty days in utero, and that lamb was finally born as Dolly.

Suppose we have a high success rate of bringing cloned human embryos to term. What kinds of developmental abnormalities would be acceptable? Acceptable to whom? Once again, the moral problems said to be raised by cloning are not unique to that technology. Every form of reproductive technology raises issues of lives worth living, of the stage at which an embryo is thought of as human, as having rights including the juridical right to state protection. Even that most benign and widespread prenatal intervention, amniocentesis, has a non-negligible risk of damaging the fetus. By concentrating on the acceptability of cloning, the commission again tried to finesse the much wider issues.

They may have done so, however, at the peril of legitimating questions about abortion and reproductive technology that the state has tried to avoid, questions raised from a religious standpoint. Despite the secular basis of the American polity, religious forces have over and over played an important role in influencing state policy. Churches and religious institutions were leading actors in the abolitionist movement and the Underground Railroad,[6] the modern civil rights movement and the resistance to the war in Vietnam. In these instances religious forces were part of, and in the case of the civil rights movement leaders of, wider social movements intervening on the side of the oppressed against then-reigning state policy. They were both liberatory and representative of a widespread sentiment that did not ultimately depend upon religious claims.

The present movements of religious forces to intervene in issues of sex, family structure, reproductive behavior, and abortion are of a different character. They are perceived by many people, both secular and religious, not as liberatory but as restrictive, not as intervening on the side of the wretched of the earth but as themselves oppressive of the widespread desire for individual autonomy. They seem to threaten the stable accommodation between Church and State that has characterized American social history. The structure of the commission's report reflects this current tension in the formation of public policy. There are two separate chapters on the moral debate, one labelled "Ethical Considerations" and the other "Religious Perspectives." By giving a separate and identifiable voice to explicitly religious views the commission has legitimated religious conviction as a front on which the issues of sex, reproduction, the definition of the family, and the status of fertilized eggs and fetuses are to be fought.

The distinction made by the commission between "religious *perspectives*" and "ethical *considerations*" is precisely the distinction between theological hermeneutics—interpretation of sacred texts—and philosophical inquiry. The religious problem is to recognize God's truth. If a natural family were defined as one man, one woman, and such children as they have produced through loving procreation; if a human life, imbued by God with a soul, is definitively initiated at conception; if sex, love, and the begetting of children are by revelation morally inseparable; then the work of bioethics commissions becomes a great deal easier. Of course, the theologians who testified were not in agreement with each other on the relevant matters, in part because they depend on different sources of revelation and in part because the meaning of those sources is not unambiguous. So some theologians, including Roman Catholics, took human beings to be "stewards" of a fixed creation, gardeners tending what has already been planted. Others, notably Jewish and Islamic scholars, emphasized a "partnership" with God that includes improving on creation. One Islamic authority thought that there was a positive imperative to intervene in the works of nature, including early embryonic development, for the sake of health.

Some Protestant commentators saw humans as "co-creators" with God and so certainly not barred from improving on present nature. In the end, some religious scholars thought cloning was definitively to be prohibited, while others thought it could be justified under some circumstances. As far as one can tell, fundamentalist Protestants were not consulted, an omission that rather weakens the usefulness of the proceedings for setting public policy. The failure to engage directly the politically most active and powerful American religious constituency, while soliciting opinions from a much safer group of "religious scholars," can only be understood as a tactic of defense of an avowedly secular state against pressure for a yet greater role for religion. Perhaps the commission was already certain of what Pat Robertson would say.

The immense strength of a religious viewpoint is that it is capable of abolishing hard ethical problems if only we can correctly decipher the meaning of what has been revealed to us.[7] It is a question of having the correct "perspective." Philosophical "considerations" are quite another matter. The painful tensions and contradictions that seem to the secular moral philosopher to be unresolvable in principle, but that demand de facto resolution in public and private action, did not appear in the testimony of any of the theologians. While they disagreed with one another, they did not have to cope with internal contradictions in their own positions. That, of course, is a great attraction of the religious perspective. It is not only poetry that tempts us to a willing suspension of disbelief.

----

1. Report of the specially created Advisory Committee on Human Radiation Experiments (October 3, 1995).

2. In fact, identical twins are genetically *more* identical than a cloned organism is to its donor. All the biologically inherited information is not carried in the genes of a cell's nucleus. A very small number of genes, sixty out of a total of 100,000 or so, are carried by intracellular bodies, the mitochondria. These mitochondrial genes specify certain essential enzyme proteins, and defects in these genes can lead to a variety of disorders. The importance of this point for cloning is that the egg cell that has had its nucleus removed to make way for the genes of the donor cell has not had its mitochondria removed. The results of the cell fusion that will give rise to the cloned embryo is then a mixture of mitochondrial genes from the donor and the recipient. Thus, it is not, strictly speaking, a perfect genetic clone of the donor organism. Identical twins, however, *are* the result of the splitting of a fertilized egg and have the same mitochondria as well as the same nucleus.

3. G. J. Annas, "Scientific discoveries and cloning: Challenges for public policy," testimony of March 12, 1997.

4. There is always the possibility, of course, that gene mutations have predisposed the child to leukemia, in which case the transplant from a genetic clone only propagates the defect.

5. It has recently been announced that success in cloning in cows is almost at hand, but by an indirect method that, if applied in humans, raises the following ethical problem. The method involves cloning embryos from adult cells, but then breaking up the embryos to use their cells for a second round of cloning. No cloned calf was yet born as of August 1, but ten are well established *en ventre leurs mères*. Even if they all reach an unimpaired adulthood, they will owe their lives to many destroyed embryos.

6. An example was the resistance to the Fugitive Slave Acts by the pious Presbyterians of Oberlin, Ohio, an excellent account of which may be found in Nat Brandt, *The Town that Started the Civil War* (Syracuse University Press, 1990).

7. Once, impelled by a love of contradiction, I asked a friend learned in the Talmud whether meat from a cow into which a single pig gene had been genetically engineered would be kosher. His reply was that the problem would not arise for the laws of *kashruth* because to make any mixed animal was already a prohibited thing.

OUR CHILDREN, OUR SELVES: THE MEANING OF CLONING FOR GAY PEOPLE

TIMOTHY F. MURPHY

Timothy F. Murphy is associate professor of philosophy in the Biomedical Sciences and head of the Medical Humanities Program, University of Illinois College of Medicine. He wrote Gay Science: The Ethics of Sexual Orientation Research *and* Ethics in an Epidemic: AIDS, Morality, and Culture. *He has edited or coedited* Writing AIDS: Gay Literature, Language, and Analysis; Justice *and the* Human Genome Project; Gay Ethics: Controversies in Outing, Civil Rights, and Sexual Science.

In this piece, Murphy argues that although gay people have embraced human cloning as a way to have children, it would be better if there were not prejudicial barriers against people adopting, parenting, and conceiving children in more traditional ways. Should such barriers be overcome, Murphy predicts, the enthusiasm of gay people for cloning will diminish.

GAY AND LESBIAN COMMENTARY ON CLONING

Some gay people reacted to the cloning of a sheep by founding Cloning Rights United Front (CRUF). This group took to the streets to protest legislation in

New York that would criminalize efforts at human cloning. Singing "Hello, Dolly," CRUF carried out small protests in Manhattan, carrying placards picturing a sheep standing on a cloud, over the caption "The Dolly Lama: Our New Spiritual Leader." Another placard proclaimed "Anti-cloning Zealotry = Homophobia."[1] CRUF founder, Randolfe Wicker, said extravagantly that the cloning of a sheep has made heterosexuality "historically obsolete." He also said that "Government has no right to keep us from reproducing ourselves." "It is reserved to each and every citizen to decide if, how, and when to reproduce." He has also said, "My twin has a right to be born." "My DNA is too precious to let die."[2] Frank Kameny, the Moses of the gay movement in the United States declared, "Keep the government off our bodies!" "Our DNA belongs only to us." "Let the cloning begin," he intoned.[3]

A lesbian columnist, Ann Northrop, has said that "in a time when we're afraid that discovery of a genetic basis [for homosexuality] would lead to people aborting us, cloning would be a way of surviving." Men and women alike could share that perspective. On the other hand, Northrop also went on to say that cloning represents a shift of volcanic proportions: ". . . the idea that women could take total control of reproduction is just stunningly exciting to me."[4] Elsewhere Northrop said: "Men are now totally irrelevant if this [human cloning] is in fact true and possible and becomes routine. Men are going to have a very hard time justifying their existence on the planet."[5] "While women might go so far as to refuse to replicate men at all, which would be an interesting concept, at the very least it would change the balance of power somewhat."[6] When criticized for these latter remarks, Northrop said that she had intended them as humorous. Some commentators nevertheless all-too-humorlessly interpreted these remarks as a declaration of homicidal intent—in the etymologically literal meaning of homicidal: man-killing.

Not to be outdone in rhetorical extravagance, Frank Kameny retorted that it will only be a matter of time before women are irrelevant too. He predicted that there would come a time when a somatic cell could by itself—i.e., without the involvement of a female ovum—be stimulated into embryonic development (let me call this somatic cell embryogenesis).[7] He also foretold the day of embryonic and fetal development in vitro, rendering women unnecessary in reproduction. One commentator has even speculated that it might become possible to join two sperm or two ova and stimulate them into embryonic development (let me call this same-sex gamete fusion embryogenesis).[8] If these visions prevailed, males and females would find in cloning another mechanism by which to go separate and reproductively independent ways.

Some commentators also made mention of the use of human cloning as a way of testing hypotheses about the biological (by which they usually mean genetic) basis of sexual orientation. They suggested that researchers could study the extent to which clones shared their progenitor's sexual orientation. Such study, it is thought, would thereby help confirm or disconfirm the "biological" nature of sexual orientation. Cloning is thus seen as a logical extension of sexual orientation research.

CLONES AS (GAY) CHILDREN

There are two main reasons why gay advocates have taken up the cause of cloning children.

Reproductive Options

One reason is that cloning seems to offer a way in which gay men and lesbians could have children in ways their own sexual relationships do not permit. The first thing to be noticed is that this mechanism—understood simply as a technique for having children—would offer few practical advantages over other reproductive techniques that already exist. For example, even if a gay man could clone himself or his partner in order to have a child, that technique would still require ovum donation, embryo transfer (ET), and a surrogate mother. Cloning would add another layer of complexity to any reproductive quest. Technicians would have to take steps to identify the appropriate somatic cells for use, undo their cellular differentiation, transfer them into an appropriately denucleated ovum, and only then transplant them into a surrogate mother's uterus, and there would be a failure rate, the best efforts notwithstanding. In effect, having a child through cloning would be more burdensome than having a child through any of the other techniques already available. And the technique could still fail if, for example, the embryo did not implant in the uterus.

Another complication of cloning lies in exactly the opposite direction. In carrying out ET, fertility clinics typically implant more than one embryo in order to increase the odds of successful implantation. It stands to reason that fertility clinics would do something like that with cloned embryos. If so, the odds of cloned twins, triplets, and even quadruplets would be substantially increased. Expecting a single cloned child/twin, adults might find themselves a triplet, quadruplet, or quintuplet as part of their quest to have children.

Even if cloning of human beings were possible, it does not follow that it would automatically be widely available. For example, it might turn out that cloning in human beings would have a very low success rate. If so, it would make sense to sort out what policy standards ought to prevail in terms of making the technique available. Health insurance companies, for example, might reasonably decline to pay for cloning services as a matter of policy if success rates were unacceptably low. In any case, it should not be assumed that some degree of success in cloning would translate into uniformly successful or widely available techniques. For these reasons, cloning will be no panacea for childless gay people wanting to be parents.

It is also unclear that gay people ought to be lining up to be the first to subject their children/twins to the unknown risks of cloning. One way to distribute experimental risk is to ask that those persons best situated to bear those risks assume them first.[9] That is, people who are vulnerable in one way or another, people who might be exploited for the gain of others who face no risks whatsoever, ought not be the first volunteers for experimental procedures. I propose that in some regards, gay people are less well situated than others to assume the risks of cloning experiments because they do not share the same liberty and equalities as the most privileged members of society. For example, gay people do not have equal rights in regard to marital status, children, reproductive technologies, employment, housing, service in the military, and so on.

Against such a background, it is worth wondering to what extent gay enthusiasm for cloning might be an artifact of obstructions society places in the way of gay people trying to have and raise children in other ways. In other words, if gay people had equivalent access to any and all mechanisms of having children in legally cognizable ways—it is unclear that they would aspire to cloning as a way of having a child/twin. Or, at least, we do not know that they would.

What I am suggesting is that the real and perceived lack of access to other means of having children can make cloning attractive as an adverse preference, a preference that except for social burdens would not be otherwise valued and pursued. I certainly do not want to rule out gay people trying to clone themselves—if the state of cloning science seemed equal to the task and meaningful reasons for doing so were forthcoming—but I do believe that the existence of impaired liberties suggests that others ought to bear most of the first risks of cloning, others whose reproductive choices are not similarly impinged, who would be making less constrained choices.

Safeguarding the Future of Gay People

There is a widespread expectation that science is closing in rapidly on factors that determine sexual orientation.[10] For many gay commentators, this is no disinterested and benign effort; on the contrary it is a commonplace of gay opinion that science may prove detrimental to the social interests of gay people by producing therapies for unwanted sexual orientations in adults and prenatal interventions capable of controlling sexual orientation in children. Rightly or wrongly, the language of genocide appears frequently in gay commentary about sexual orientation research. This is not surprising as minorities are often suspicious of science when science has been the source of individious classifications and stereotypes drawn up at their expense. Medicine's interpretation of homosexuality as pathological and its presumptive willingness to eliminate that pathology loom large as the historical foundations of the suspicion gay people harbor toward science.

In light of this social history, it is ironic that gay people look to cloning as a way of ensuring the continuation of gay people in kind. That is, some commentators look to cloning precisely as way in which gay people could offset losses to their numbers. Cloning would thereby serve a protective function by helping keep the number of gay people high enough to maintain an effective social and political community. In fact, however, there is little evidence that warrants the assumption that a cloned human being would have the same sexual orientation as his or her progenitor. On the basis of the available evidence, it simply does not follow that having the same genome will guarantee the same sexual orientation in a clone. *Every* relevant study of twins has shown, for example, that monozygotic twins do not always share the same sexual orientation.[11] There is no reason to think, therefore, that a clone would necessarily share its progenitor's sexual orientation. As John Money has never ceased pointing out, it is nature and nurture working in tandem to produce sexual interests, never nature nor nurture alone.[12]

Let me make this point more plainly by way of an example. Let us assume for the sake of the argument that psychologist Joseph Nicolosi's hypothesis about the origins of homosexual orientation in males is true. Nicolosi has argued that males become homosexual because at a very early age they experience a psychic injury at the hands of their fathers.[13] It might very well prove that a gay male raising his cloned child/twin would never inflict such an injury on that child—precisely because he has taken pains to avoid injuring the child in ways he was himself injured. Consequently, because the gay father inflicts no psychic injury on the potentially gay child/twin, the child's sexual orientation does not emerge as homosexual—even if there were some underlying biological disposition toward that outcome. Having a gay parent willing to protect a child from the slings and arrows of outrageous homophobic fortune might well defuse the very developmental experiences that play a role in determining sexual orientation. This is not to say that Nicolosi's theory is true (indeed, there are many reasons why I think it is defectively argued),[14] but it is to say that there is no necessary reason to think that clones would necessarily share the same sexual orientation as their progenitors even if they share common genomes. It may be that within each genome there are several potentials with respect to sexual interests, and that these possibilities are expressed for a combination of reasons: fetal nutrition, fetal hormone exposure, wired-in dispositions in learning sexual behavior, and developmental experiences, for example.

Genetics—and cloning—need not be sexual orientation destiny in any simple way.

None of this is to say that cloning would be irrelevant to the study of sexual orientation. Yet as a matter of science, this line of inquiry would have its own limitations. For example, suppose that a gay man cloned himself and that he firmly believed—as many gay people do—that sexual orientation is determined by genetic endowment. In expecting the child to be gay, the father/brother may very well raise the child in conformity to that expectation: encouraging him to admire boys, protecting him against social assumptions that he will be heterosexual, and so on. In short, this kind of rearing might have the effect of a self-fulfilling prophecy. If so, researchers would be blinded by social influences to the role genetics plays in sexual orientation. Certainly, it must be said, a study of the commonality of sexual orientation between child/twin and progenitor would be interesting, but this might not yield the unequivocal results expected of it.

Even if we knew that clones usually shared the same sexual orientation (for whatever causal reasons), it is unclear how many gay children could be produced by cloning. Given the comparative effort required to clone and given obstacles in access, it is unlikely that gay people would use cloning on any large scale. If they did, they would still have both gay *and* straight children/twins because of potential for variation. It would remain to be seen whether the numbers of gay people produced after taking these effects into account would be sufficient to offset the losses of gay people if—simultaneously—people used sexual orientation therapies to rid themselves of unwanted homosexual interests and prenatal interventions to avoid homosexual orientations in their children. This is at present no more than speculation, but it does not seem unreasonable to believe that if there were techniques offering parents control over the sexual orientation of themselves or their children, these techniques would be more widely used than cloning techniques to ensure that one's children/twins were gay. To put it another way, a vast number of gay people would have to use cloning and have gay children/twins to make meaningful inroads against the feared use of future sexual orientation technologies. Moreover, it is to be remembered that the human clones will—presumably—need to be gestated in the uterus of a human female. I do not know what number of surrogate mothers would be willing to become pregnant for the express purpose of bearing a clone of a gay male, but I would guess that the number would be small. Lesbians would not face this problem as long as they were able and wanted to bear their clonal pregnancies themselves. In any case, the willingness of women to bear clones for gay people would be one more factor limiting widespread use of cloning. It is hard to see, therefore, that gay people should look to cloning as a remedy for a problem of falling numbers.

In view of the foregoing limitations, I do not think that cloning has the capacity to serve as a major social engine for gay issues. Rather than thinking that the commentators whom I quote at the onset to be merely media opportunists, I want to say that I interpret advocacy of cloning by gay people as a moral displacement. That is, I interpret calls for the cloning of (gay and lesbian) human beings not literally but figuratively, as expressing worry about the social standing of gay men and lesbians at present, not in the future.

Gay people do not have the same standing in relationship to children as others do, and they worry that future technologies and social currents will worsen that already tenuous relation. A Virginia court found, for example, a woman to be unfit to have custody of *her own child* because the woman was in a lesbian relationship.[15] A Minnesota court found, in another case, that lesbians did not have the same right to reproductive services as other women, and it did so *despite* the fact that Minnesota has a statute that forbids discrimination on the basis of sexual orientation.[16] These kinds of inequalities could be multiplied without much effort. Gay and lesbian advocacy of cloning might therefore be seen as much as a protest of current inequality as a claim on future technologies. In advocating cloning, commentators are asserting a principle of equality they wish to see respected even if most gay men and lesbians would never use cloning.

CONCLUSION

Apart from the views expressed by self-appointed cloning activists, there are no scientifically meaningful polls of how gay people view the desirability of cloning. It almost goes without saying that neither are there studies of how gay people would actually avail themselves of cloning techniques were they available. Given this informational vacuum, no assumption should be made that gay people stand rank and file behind their media leaders. Whatever support there might be, I suggest that it is probably better to interpret gay advocacy of cloning less as a literal credo of political entitlement than as a measure of social insecurity. If gay people had more untroubled access to current reproductive technologies, foster care, and adoption, and could not be stripped of legal rights to their children, I doubt that we would see much gay championing of cloning, somatic cell embryogenesis, parthenogenesis, same-sex gamete fusion embryogenesis, and other novel forms of embryogenesis. To see cloning as something essential to the future of gay people and gay parenting is, I think, to misapprehend the solution to the problems that beset gay people.

Nothing I have said here, however, amounts to an argument that no human being should ever be cloned or that no gay man or lesbian should ever produce a child this way. I merely mean to say that the state of mammalian cloning science is insufficiently developed to see it as a valuable reproductive option for gay men and lesbians, to claim it as a right, or to see it as an ersatz solution for the injustices gay people already suffer in reproductive choices and elsewhere. Neither is the state of knowledge about effects of cloning *on children* anywhere near the kind of threshold that would justify claiming the technique in the name of gay or feminist activism. Were cloning to prove safe in the production *and* rearing of children, I certainly do think gay people ought to have access to what techniques become available. Until that time, however, there are many causes more important to gay people waiting for attention, causes more important than being able to use adult somatic cells to produce twins to be raised as children. Opening doors closed to gay men and lesbians in regard to adoption and IVF/ET services, for example, would be a good first step toward improving the social standing of gay people and reducing the extent to which cloning looks like a valuable solution to problems of social justice.

---

1. Christopher Rapp, "Gay Clones," *GayToday* [http://gaytoday.badpuppy.com/viewpoint/index.htm], May 27, 1997.

2. In an explanatory letter, Wicker reined in his statement, saying: "In the middle of the night, being interviewed by an old friend, perhaps slightly impaired by alcohol, I confess to becoming so enamored with my verbal abilities (i.e., so carried away with hearing myself talk cleverly) that I gave in to the urge for rhetorical excess and did in fact say: 'Heterosexuality as a route to reproduction is now historically obsolete.' Factually, it is a stupid and inaccurate statement which has haunted me ever since. Heterosexuality will never become 'historically obsolete' and will doubtlessly be the predominant mode of reproduction as long as the majority of men and women seethe with sexual desire for one another. Given the opportunity, I've restated the essential and important idea buried under that inflammatory rhetoric: 'Cloning renders heterosexuality's historic monopoly on reproduction obsolete.'" Randolfe Wicker, "Letter to Heterodoxy Magazine," *GayToday*, May 27, 1997.

3. Jorget Harper, "Our Sheepish Revolution," *Outlines*, April 1997, pp. 32–33.

4. Harper, "Our Sheepish Revolution."

5. Jack Nicols, "'Support Cloning' Say Top-Name Lesbian & Gay Activists," *GayToday*, Mar. 3, 1997.

6. Harper, "Our Sheepish Revolution."

7. Simon LeVay has speculated that it will be possible someday to carry out parthenogenesis in human beings, i.e., taking female or male gametes and treating them so that they begin embryonic development. "Queer Science," *Gay Chicago*, Nov. 18, 1996.

8. Chandler Burr has said as much, saying that two men could join half their genes and produce a child thereby. Rapp, "Gay Clones."

9. Hans Jonas, *Philosophical Essays: From Ancient Creed to Technological Man* (Chicago: University of Chicago Press, 1988).

10. Timothy F. Murphy, *Gay Science: The Ethics of Sexual Orientation Research* (New York: Columbia University Press, 1997), pp. 24–41.

11. Simon LeVay discusses these studies in *Queer Science: The Use and Abuse of Research on Homosexuality* (Cambridge: MIT, 1996), pp. 173–178.

12. John Money, *Gay, Straight, and In-Between: The Sexology of Erotic Orientation* (New York: Oxford University Press, 1988).

13. Joseph Nicolosi, *The Reparative Therapy of Male Homosexuality* (Northvale, N.J.: Jason Aronson, 1991).

14. Murphy, *Gay Science*, pp. 37–41.

15. Tziva Gover, "Fighting for Our Children," *The Advocate*, NOvember 26, 1997, pp. 22–30.

16. "In the Courts," *Hastings Center Report* 26 (1996): 47.

INDEX

ABOUT THE EDITOR

Gregory E. Pence is professor of philosophy in the Schools of Medicine and Arts/Humanities at the University of Alabama, Birmingham, where he has taught and written about bioethics for over twenty years. He previously published in the *New York Times, Newsweek,* the *Wall Street Journal,* and the *Journal of the American Medical Association.* He is author of *Who's Afraid of Human Cloning?* (Rowman & Littlefield, 1998) and *Classic Cases in Medical Ethics* (2nd ed., McGraw-Hill, 1995), editor of *Classic Works in Medical Ethics* (McGraw-Hill, 1997), and coauthor of *Seven Dilemmas in World Religions* (Paragon House, 1995).